PERGAMON INTERNATIONAL LIBRARY
of Science, Technology, Engineering and Social Studies
*The 1000-volume original paperback library in aid of education,
industrial training and the enjoyment of leisure*
Publisher: Robert Maxwell, M.C.

SCALE MODELS IN ENGINEERING

Fundamentals and Applications

THE PERGAMON TEXTBOOK
INSPECTION COPY SERVICE

An inspection copy of any book published in the Pergamon International Library
will gladly be sent to academic staff without obligation for their consideration for
course adoption or recommendation. Copies may be retained for a period of 60 days
from receipt and returned if not suitable. When a particular title is adopted or
recommended for adoption for class use and the recommendation results in a sale
of 12 or more copies the inspection copy may be retained with our compliments.
The Publishers will be pleased to receive suggestions for revised editions and new
titles to be published in this important international Library.

SCALE MODELS IN ENGINEERING

Fundamentals and Applications

DIETERICH J. SCHURING

CALSPAN CORPORATION, BUFFALO, N.Y.
FORMERLY CORNELL AERONAUTICAL LABORATORY

PERGAMON PRESS

OXFORD · NEW YORK · TORONTO · SYDNEY · PARIS · FRANKFURT

U.K.	Pergamon Press Ltd., Headington Hill Hall, Oxford OX3 0BW, England
U.S.A.	Pergamon Press Inc., Maxwell House, Fairview Park, Elmsford, New York 10523, U.S.A.
CANADA	Pergamon of Canada Ltd., 75 The East Mall, Toronto, Ontario, Canada
AUSTRALIA	Pergamon Press (Aust.) Pty. Ltd., 19a Boundary Street, Rushcutters Bay, N.S.W. 2011, Australia
FRANCE	Pergamon Press SARL, 24 rue des Ecoles, 75240 Paris, Cedex 05, France
WEST GERMANY	Pergamon Press GmbH, 6242 Kronberg-Taunus, Pferdstrasse 1, Frankfurt-am-Main, West Germany

First edition 1977

Library of Congress Cataloging in Publication Data

Schuring, Dieterich J
Scale models in engineering.

Bibliography: p.
Includes index.
1. Engineering models. I. Title.
TA177-S38 1977   620'.0022'8   76-50085
ISBN 0-08-020861-4 hard.
ISBN 0-08-020860-6 pbk.

*In order to make this volume available as economically and rapidly as possible the
author's typescript has been reproduced in its original form. This method unfortunately
has its typographical limitations but it is hoped that they in no way distract the reader.*

Printed in Great Britain by A. Wheaton & Co., Exeter

to my wife,

Irene

CONTENTS

PART II

APPLICATIONS

PART II
APPLICATIONS (Cont.)

PREFACE

Despite the current strong tendency toward computerization, the researcher in physical and engineering sciences relies more than ever on experimentation. More theory demands more testing, not less; otherwise, the theorist would drift toward playing mere games.

Experiments are often planned and profitably executed as scale model experiments; i.e., lengths, times, forces, temperatures, and electric currents of the original system, or prototype, are scaled such that the resulting system, or scale model, can be studied as a valid substitute for the original. Such experiments accomplish three salutary things: (a) they permit the transformation to manageable proportions of certain features of original phenomena that are difficult to handle, such as extreme size, very slow flow, vast energy release, and microscopic dimensions; (b) they shorten experimentation by compacting the number of variables; and (c) they promote a deeper understanding of the phenomena. The last point is frequently overlooked, perhaps because of exaggerated reliance on the formalism of the pi-theorem.

The term scale model testing usually evokes images of miniature hardware or suggests simplifications of a complex system -- connotations that fail to convey the breadth and depth of this challenging and fruitful line of experimental research. Scale models are defined to include length scale factors of any magnitude so that the experimental hardware can be made smaller, equal, or larger than the original configuration. In addition, the term implies true representation of an original phenomenon without simplification or distortion, although this goal is not always easy to accomplish. It indicates, furthermore, genuine experimentation, the source of new insight into the real world and of guidance in improving theories and engineering hardware.

From the author's long-standing interest in scale model experiments and his work has evolved a simple and fundamental method of designing scale model experiments. In this novel method, which he calls the law approach, pi-numbers and model rules are extracted directly from the physical laws

governing the phenomenon under study. Having applied the law approach for
many years, the author believes that it is not only an expedient and simple
way to collect all necessary pi-numbers but also, and more importantly, an
effective method to understand fully the problem to be investigated. Furthermore, the new approach prepares the ground for one of the most difficult but
never systematically explored tasks in scale modeling; namely, to resolve
conflicting modeling claims by proper relaxations.

Chapter 1 of Part I explores the background of scale modeling. The design
procedure of scale models and experiments is explained in Chapter 2, after a
discussion of formal principles. Chapter 3 analyses the relaxation methods
commonly applied to conflicting requirements in model design. Although these
methods are often referred to as distortion methods, the author prefers the
term relaxation, which correctly denotes a lessening of rigor rather than
an unnatural twisting out of shape, as suggested by the term distortion.

Part II is devoted to case studies. Since scale model experimentation is,
like all experimentation, as much an art as a science, the author feels that
close contact with practical problems is important for a true comprehension
of general principles. The case studies are selected from modern fields of
model application; all have been interpreted uniformly.

An appendix presents a number of problems and a catalog of classical pi-
numbers and their physical background.

This book is designed not only as a college textbook for senior and graduate
levels but also as a working reference for practicing engineers. Since the
material was prepared originally for an extension course at the University of
California, Los Angeles, the book will be particularly suitable for such
courses.

Sincere thanks are due to Dr. R.I. Emori who was a collaborator of the book
in its earlier stages before he returned to Japan in 1972. His efforts and
skills have greatly contributed to the final form of this volume.

PART 1

FUNDAMENTALS

CHAPTER 1

INTRODUCTION

<u>Why Scale Model Experiments</u>

The word "model" carries many broad implications. Outstanding students, futuristic homes, new cars, attractive young women -- these as well as ideas, images, concepts, analogies, and mathematical descriptions are all frequently referred to as models. Quite contrary to this rather liberal practice, the models discussed in this book are well defined: they involve pieces of hardware proportioned after certain "original" systems, or prototypes, in most or all of their important variables and constants -- their lengths, velocities, forces, densities, viscosities, etc. Thus, through "scaling" of all influential quantities, the original phenomenon, or prototype, is transformed into a "scale" model. That is, the phenomenon is transformed into a similar system which preserves the relative values and proportions of the prototype, even though it may require less (more) space, proceed faster (slower), involve smaller (larger) forces, or result in lower (higher) temperatures. Note that the scaling applies not only to linear dimensions but to all other important quantities, such as time, force, density, and viscosity.

Engineers and scientists have been profitably using scale models for many years. A.L. Cauchy, a French mathematician, investigated models of vibrating rods and plates as early as 1829. W. Froude made the first water-basin model for designing watercraft in 1869. O. Reynolds published his classic model experiments on fluid motion in pipes in 1883. And not many years later, the Wright brothers built a wind tunnel to test wing models.

The scaled reproduction of a physical phenomenon or an engineering system can be advantageous for four reasons. The first applies to all experimental work: the problem at hand is too complex or too little explored to be amenable to an analytical solution; empirical information is needed. The other three reasons apply to scale modeling proper: scale models permit transformation to manageable proportions of systems that, like a suspension bridge, may be too large for direct experimentation; or, like a spacecraft,

inaccessible; or, like a firestorm, unmanageable; or, like seepage, too
slow to work with. This reason is often presented as the overriding rationale
for resorting to scale model work. But two other reasons are not less important: scale modeling shortens experimentation, as we will see, and it
promotes (in fact, requires) a deeper understanding of the phenomenon under
investigation.

Of these reasons, the last is perhaps the most difficult to grasp. No
physical phenomenon can be modeled without a preliminary analysis of its
inner mechanism. The preparatory analysis, i.e., the formulation of a
hypothesis as to which physical laws govern a phenomenon and which can be
neglected, is the most difficult and the most challenging part of scale
modeling; it is also the most rewarding. For this reason, scale model
experiments are often called an art rather than a technique. The process
of modeling cannot be logically derived from a few axioms; it must be found
anew for each problem.

As an example, consider the problem of predicting the trajectories of two
automobiles involved in a collision. Let an experimental investigation of
all trajectories of front-, side-, and rear-impact collisions of full-size
cars at various speeds be ruled out as too expensive. An attractive alter-
native might be to experiment with small car models that, when operated
according to some rules, would yield accurate predictions. From results of
real accidents, we may find that a large portion of the vehicle body remains
intact in a collision; and from this we may deduce that up to a certain
speed, say 100 km/h, except in the immediate vicinity of impact, the vehicles
can be considered rigid bodies. Could we, then, in a scale model experiment,
use rigid models -- for instance, rigid blocks? This appears most unlikely
because the peripheral crushing of automobiles in collisions softens the
impact considerably. The question is then -- how to simulate the crusha-
bility of the body in a model without unduly complicating the manufacturing.
There are other questions, too. Can we extract the necessary information
from a model automobile that is confined to only plane motions? Or must we
have a third degree of freedom in motion? How to account for the deflections
of suspensions and tires during collision? How to recognize the friction
between the two sliding vehicle bodies in a collision? The answers cannot
be derived solely from logical processes; instead, they must be extracted

from our understanding of the phenomena and from our experience. How few
misinterpretations we make depends not so much on rules and techniques as
on our skill to penetrate the phenomenon in depth. Hence, this book dwells
as little as possible on formal approaches. Instead, it demonstrates the
principles of scale model experimentation with practical case studies, taken
from the literature and from experiments by the author. He believes that,
like many other ideas, the idea of scale modeling can be grasped more
easily by working from real life examples than by studying abstractions.

To circumscribe the scope of this book in still another -- negative -- way,
we will touch briefly upon some types of models that, although fitting the
general description of a model as representing something not readily available, fall outside the precise definition of scale models.

Subjective models, such as the conceptual model developed by the philosopher,
or the sociologist, reflect his views of the structure of human nature and
society. Subjective models can also demonstrate ideas on our future urban
environment, such as the architect's miniature model of a city. Although
these models are powerful in their quest for representation, they range
outside the interest of our book because they lack the exactness and strength
of scientific research. They are experienced and judged by individuals, not
by objective measures.

A second group of models -- the *qualitative models* -- comes closer to the
desired specification. They skirt or even penetrate the domain of precise
numbers and functional relations without, however, leaving the qualitative
domain. Most of them are an aid in deliniating the boundaries of new
engineering devices rather than in pinpointing exact data. Consider for
example:

- The breadboard model, which has little resemblance to the
  final article, but is a physical help in ensuring proper
  functioning of a new device;

- the mock-up model, which reflects the external appearance
  of a new concept but lacks proper functioning;

- the test bed, the pilot plant, and the development model,
  which are first assemblies of essential elements of new
  machinery, serving to recognize malfunctions and to direct
  further developments; and

- the prototype, which is the end product of the development
  stage, allowing final adjustments and initiating the first
  production series.

In the same group belong models designed to give a rough, qualitative idea
about the performance of a new technical design. Some examples are:

- Geometrically scaled paperboard models of frame structures
  that aid the designer to assess stiffness and elastic deformations under load;

- scale models of suspension bridges exposed to smoke streamers
  so as to visualize vortex patterns of wind around the girders;
  and

- aluminum-powder streak photographs of water motions around small
  models of underwater structures.

Testing these models resembles experimental work insofar as it is directed
toward understanding new machinery or a new process. The tests guide further
improvements, help to avoid mistakes at an early stage of development, and
ensure that the specifications of the desired end product are met. Because
precision is not required, however, these models are a qualitative help
rather than a quantitative tool of scientific inquiry. They are not subjects
of this book.

Another group of models called *analogs* are designed to demonstrate known
quantitative relations among governing parameters. Toy models of cars,
ships, and airplanes are simple examples. A geographic map showing previously
measured terrain features would be another example, or a diagram displaying
the previously measured interplay among variables. M.G. Murphy (Ref. at the
end of Part I) lists and describes a large number of analog models; for
instance, the famous soap-film membrane exhibiting the torsional stress
distribution of an elastic member, or the diffusion of a solution in a test
tube simulating the heat flow in a rod. The mechano-electrical analog and

the thermo-electrical analog based on the identity of mathematical equations describing mechanical, thermal, and electrical processes are other well known examples. Most of these analog models have been replaced by computer models.

Since analog models parallel a phenomenon by using already established functional relations rather than discover unknown relations, they are treated in this book only marginally.

Last, a group of models distinguished from our scale models are the *mathematical* models. With the advent of computers and system analysis, the use of mathematical models has been spreading into all scientific areas, even those that are not thoroughly amenable to quantitative treatment, such as human behavior, mental processes, biological functions, urban planning, and management. Mathematical models, or computer models as they are often called, now span the whole range of scientific inquiry from purely phenomenological descriptions to general theories fusing a vast amount of data into simple relations.

A mathematical model can be formulated only when a number of detailed inputs are furnished, such as firm experimental evidence, good approximations, and reasonable simplifications. Since our attention here is directed toward acquiring these inputs rather than formulizing them, mathematical models are treated only occasionally in this book.

To summarize, subjective models, qualitative models, analog models, and mathematical models are not subject of this book. The models with which we are concerned are *scale* models, that is, experimental models structured to mirror the true physical behavior of an original phenomenon, or a prototype. Scale models are valid substitutes for systems that for some reason (too large, too small, too slow, too fast, too expensive, inaccessible, unmanageable) cannot be explored on the prototype level. If scaled correctly, deflections, deformations, speeds, forces, accelerations, energies, temperatures, electric currents, magnetic fields, and a host of other relevant quantities measured on the scale model permit prediction of the corresponding quantities of the prototype design.

Scale model tests cannot be performed without some insight into the given
phenomenon, an insight that will grow into an understanding of the basic
structure of the system at hand as the scale model tests progress. Hence,
scale model experiments are uniquely qualified to help establish both the
design *and* the fundamental nature of new engineering hardware.

Fields of Application

The simplest scale models are those built by architects. Similar to these
are the miniature table models of a city section used by city planners;
viewing the models from the eye level of a miniature pedestrian, they can
obtain a realistic, three-dimensional perspective otherwise difficult to
achieve. A model of the lighting in a room is another example, one that
requires even more architectural fidelity, with the room, furniture, and
windows all to exact scale. Even the reflectivity of important surfaces is
often duplicated to reflect light from an artificial sky that is carefully
designed to simulate natural daylight.

Other elementary models are found in toy and hobby shops. Some are so true
in form and function that they can be used in simple scale model experiments.

Architectural and toy models, however high their fidelity, are judged sub-
jectively, that is, without scientific exactness, unlike the model pictured in
Fig. 1. It is a remote-controlled scale model of a lunar rover manufactured
with the highest possible accuracy. Its masses, moments of inertia, elasticity
speed, steerability, and power are all carefully scaled; operated on simulated
lunar soil, the model reflects the true motions and forces of the prototype on
the moon under lunar gravitational acceleration.

One of the oldest and best known fields of scale model experimentation is
naval architecture. For many decades, no ship of importance has been designed
without the help of a model-testing tank, or water basin. Frictional and
wave-making resistance, propeller performance, ship maneuverability in smooth
and rough water, cavitation, ship bending and vibrations due to wave impact
and slamming, seakeeping, and many other performance characteristics would
be impossible to predict without model experiments. Figure 2 shows a landing

Fig. 1.   Remotely controlled lunar roving vehicle for two astronauts and their gear. The model is 28 cm wide and 56 cm long, or six times smaller than the prototype, in accordance with the ratio of g_{moon}/g_{earth} = 1/6. The maximum speed (of model and prototype) is 8 km/h. Tests are run on crushed basalt, with grain sizes and shear characteristics scaled after lunar soil samples collected on previous moon missions.
(Photo from U.S. Army Engineer Waterways Experiment Station, Vicksburg, Miss.)

Fig. 2.   Small model of a landing craft under test in water tank. The model is towed by an
           electrically powered carriage spanning the tank and running on rails that assure con-
           stant height above still water level (accuracy, 0.1 mm). The linkage between model
           and carriage measures the towing force and permits the model to assume a natural
           attitude in pitch and heave.
           (Photo courtesy of Davidson Laboratory, Stevens Institute of Technology, Hoboken,N.J.)

craft model. Water basins are also used for testing models of other kinds of
watercraft. Figure 3 depicts a scale model of an amphibious air cushion
vehicle during a test run.

As valuable as the water basin is the wind tunnel. Capable of subsonic,
supersonic, hypersonic, and hypervelocity air flow, wind tunnels can simulate
velocities ranging from a light breeze to many times the speed of sound.
And they can accommodate scale models of structures, aircraft, and spacecraft
for studies of turbulence, drag, lift, pressures, buffeting, flutter, and
other phenomena. Figure 4 shows a supersonic aircraft model suspended in a
wind tunnel for lift and drag investigations; Fig. 5 depicts miniature models
used in wind tunnel studies of sonic-boom generation.

The wind tunnel has extended its usefulness to modeling atmospheric phenomena,
such as wind turbulence around tall buildings or distribution of pollutants.
Figure 6 shows a model of an office tower and surrounding pedestrian plaza
in an atmospheric wind tunnel.

A field of gradually diminishing importance for model application is that of
large and complex structures, such as suspension bridges, dams, frames,
shells, vessels, towers, and the like. Laboratory experiments are being
replaced with high-speed computers and advanced numerical methods. Model
experiments are still used for many structural problems, however, such as launch-
induced vibrations of giant space vehicles (Fig. 7), blast response of underground shelters, buckling of composite materials, and frequency response of
automobile bodies.

Experimental research of flow phenomena with models is not restricted to the
water basin and the wind tunnel. Another celebrated model facility is the
hydrology laboratory, where studies can be made of the hydraulic phenomena
of coast lines, estuaries, open channels, spillways, rivers, harbors, locks,
reservoirs, watersheds, tsunamis, etc. Figure 8 shows a large hydrology model
to study siltations and salinity distributions in an estuary; Fig. 9, a
model to investigate wave actions in a habor.

Designers of hydroelectric machinery also make extensive use of scale models,
for the performance of pumps, turbines, etc., cannot yet be completely

Fig. 3.   Remotely-controlled model of a 160-ton Amphibious Assault Landing Craft. The
model is 4.5 m long and runs on a 25-cm thick air cushion (length scale 1:6). It is
powered by two 22-kW gasoline engines, which drive its aft-mounted ducted propellers
and the fans for the air cushion. (Photo courtesy of Bell Aerospace Division of Tex-
tron, New Orleans, La.)

Fig. 4.   Model of supersonic aircraft installed in transonic wind tunnel. The model is mounted
on a force-sensing sting capable of measuring all six aerodynamic forces and moments.
The airflow is bleeded through perforations in the walls of the wind tunnel, to alleviate
the effects of shock waves and to minimize reflections.
(Photo courtesy of Calspan Corp., Buffalo, N.Y.)

Fig. 5.   Precision models of various aircraft used to study farfield effects of sonic booms in a
supersonic wind tunnel.
(Photo from National Aeronautics and Space Administration, Washington, D.C.
Original Source: H.W. Carlson and O.A. Morris, *AIAA J. Aircraft*, **4** [1967] 245).

Fig. 6.   Model (length scale 1:400) of the Commerce House in Seattle, Washington, tested in
           an atmospheric wind tunnel to determine the wind pressures on the cladding of the
           tower and the flow pattern in the pedestrian area.
           (Photo courtesy of Calspan Corp., Buffalo, N.Y.)

Fig. 7.   Elastic model of Saturn V used to study resonant frequencies, bending mode shapes, and damping. The model (length scale 1:5) is suspended by two vertical cables attached to a support bar at the base; the fuel tanks are filled with water; vibrations are introduced by horizontal shakers.
(Photo from National Aeronautics and Space Administration, Washington, D.C.)

Fig. 8.   Hudson River siltation model, with Manhattan Island (center) and Statue of Liberty
(foreground). The entire New York Harbor area is reproduced to a scale of 1:100
(vertically and horizontally). Tides and tidal currents are produced to scale by tide
generators (at the model boundaries). The model can be operated with salt water and
fresh water introduced downstream and upstream, respectively. Silting is studied by
adding lightweight granular material that moves and deposits under the influence of
model currents.
(Photo from U.S. Army Waterways Experiment Station, Vicksburg, Miss.)

(a) UNPROTECTED, MODEL WAVE HEIGHT 6 cm   (b) PROTECTED BY BREAKWATER

Fig. 9.   Model of proposed Noyo Harbor north of San Francisco, California. The model is
used to test the effect of various breakwater designs on storm waves which would reach
heights of 8 m. The length scale factor (vertical and horizontal) of the 600 by 600 m
harbor area (depth up to 50 m) is 1:100. Waves are generated by a wavemaking plunger.
(Photos from U.S. Army Waterways Experiment Station, Vicksburg, Miss.)

predicted by theory. To obtain empirical data and confirm their computa-
tions, especially when very large machines are being developed, they prefer
to perform experiments on a smaller scale. Three scaled pumps built for
generating empirical data are shown in Fig. 10.

Meteorologists and geophysicists are another group long interested in scale
model experiments. As early as the late eighteenth century, an attempt was
made to construct a laboratory model of a cyclone. Today, the study of geophysical phenomena in the laboratory on a miniature scale is well advanced.
One example of atmospheric modeling in wind tunnels has already been given
(see Fig. 6). A different kind of model is shown in Fig. 11: a model of
extremely slow fluid flow, a miniature glacier composed of water and kaolin.
Because geophysical models often represent an appreciable fraction of the
earth's surface, the effects of earth curvature and rotation must be taken
into account.

These are only a few examples of scale models; there is no limit to scale
model experimentation within the realm of quantifiable physical phenomena.
For instance, investigations of the spread of fire and conflagrations have
profited greatly from model experimentation. Acoustical models have contributed to the improvement of auditoriums (Fig. 12), as well as to an understanding of the attenuation of noise and sonic boom effects (Fig. 5). In
chemical engineering, many new processes are first tried on a small scale
before a full-scale plant is built. Many machines, reactors, and furnaces
can be examined and improved in scale model experiments before a final
design. Rocketry and space flight have enormously benefited from model
experiments. Figure 13 shows the model of a space shuttle vehicle in a
supersonic wind tunnel. High-speed impact phenomena constitute another field
of fruitful model application. A final example is the extensive model work
performed in soil mechanics; earthquakes, tunnels, pile foundations, and
underground explosions, as well as bulldozers, plows, tillers, and off-road
vehicles have all been investigated with models. Figure 14 shows full-scale
and model marsh vehicles; Figure 15, model and full-scale tires for a large
earthmover.

Fig. 10   Three geometrically similar hydraulic pumps, designed to verify scaling laws for
           specific work, flow rate, torque, and efficiency for a range of speeds and pressure ratios.
           (Photo courtesy of Prof. M. Shirakura, Japan)

Fig. 11.   Model glacier made from kaolin. Faults and crevasses closely resemble those of field
observations.
(Photo from *Models in Geography*, ed. by R.J. Chorley and P. Haggett, 1967.
Courtesy of Methuen & Co., London)

Fig. 12.   Acoustic model designed to study audience absorption in lecture halls. The pistol
provides an impulse sound source of a wide frequency spectrum. The sound is diffused
by panels, (partly) reflected by the model auditors (made from polyurethane foam),
picked up by a microphone, recorded, and analyzed.
(Photo from L.W. Hegvold, "A 1:8 scale model auditor," *Applied Acoustics* , 4 [1971]
237-256. Courtesy of Dr. Hegvold, Ottawa, Canada)

Fig. 13   Model designed to investigate heating in the base region of a space shuttle vehicle during the boost phase of the mission. The 1:44 scale model is mounted upside down in a supersonic wind tunnel; it is equipped with hydrogen-oxygen engines and solid propellant motors. (Photo courtesy of Calspan Corp., Buffalo, N.Y.)

Fig. 14.   Full-scale marsh vehicle and one-fourth scale model under test, to verify model rules
            for sinkage, slip, drawbar pull, and power requirements (see Case Study 5).
            (Photo courtesy of C.J. Nuttall, Jr., Vicksburg, Miss.)

Fig. 15.   Full-scale earthmover tire and one-fourth scale model designed for model tests of
            earthmoving equipment (see Case Study 6).
            (Photo courtesy of Firestone Tire and Rubber Co., Akron, O.)

This incomplete survey can only give an impression of the versatility and the possibilities of experimental investigation by scale models. Some of these are discussed as case studies in Part II of the book.

CHAPTER 2

PRINCIPLES AND DESIGN OF SCALE MODEL EXPERIMENTS

Fundamental Requirements

The general concept of scale modeling can be visualized by considering small elements of the prototype and their corresponding elements in the model.

In studying the prototype elements, we account for their physical data of interest -- geometry, pressure, stress, deformation, weight, velocity, acceleration, frequency, magnetic field strength, electric current, etc. Homologous[1] behavior of the corresponding model elements is secured if each quantity[2] of each prototype element can be transformed into the corresponding quantity of corresponding model elements through multiplication by a respective constant factor or "scale factor." If, for instance, v_1 , v_2 , v_3 , ... v_n are the velocities of, respectively, the first, second, third,...nth element of the prototype, and v_1', v_2', v_3',... v_n' the corresponding velocities of, respectively, the corresponding first, second, third,...nth element of the model, then homologous behavior with respect to velocity requires that

$$\frac{v_1}{v_1'} = \frac{v_2}{v_2'} = \frac{v_3}{v_3'} = \cdots = \frac{v_n}{v_n'} = v^*$$

where v^* is the velocity scale factor (Fig. 16).

Analogous requirements exist for all other corresponding quantities. In general,

$$\frac{q_1}{q_1'} = \frac{q_2}{q_2'} = \frac{q_3}{q_3'} = \cdots \frac{q_n}{q_n'} = q^*$$

---

[1] Homologous -- having the same relative position, proportion, value, or structure (derived from the Greek homologos -- agreeing, assenting).

[2] The term "quantity" is used throughout this book as a short notation for any quantifiable parameter, variable, constant, or product, whether scalar, vector, or tensor.

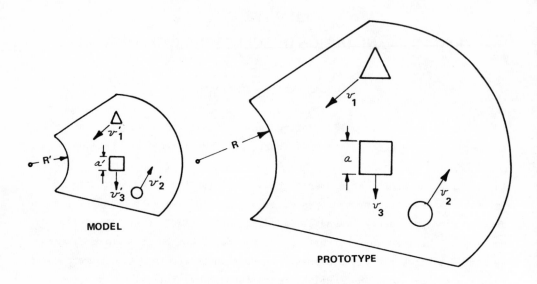

Fig. 16. Homologous behavior of three corresponding model and prototype elements with respect to velocity

$$\frac{v_1}{v_1'} = \frac{v_2}{v_2'} = \frac{v_3}{v_3'} = 2$$

All corresponding lengths are subjected to the same length scale factor. For instance, $R/R' = a/a'$.

where q_i and q_i' ($i = 1,2,3,\ldots n$) are corresponding quantities of the same kind (the primed quantities refer to the model), and q^* is the scale factor. For geometrical similarlity, q^* represents the length scale factor; for temporal similarity, the time scale factor; for similarity of forces, the force scale factor; and so forth.

As an example of homologous systems, consider two structural beams, a prototype and a model subjected to periodically changing loads (Fig. 17). If all the essential quantities of the beam are scaled correctly (only the requirements for scaling are pointed out here; how to accomplish scaling is discussed in the next chapter), i.e., if all corresponding lengths are linked by the length scale factor

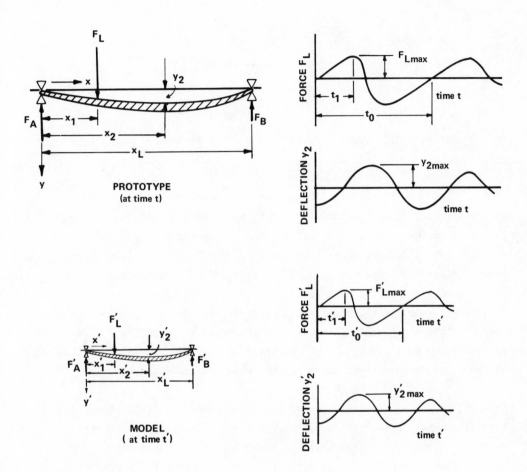

Fig. 17.   Prototype and homologous (scale) model of vibrating beam.

$$\ell^* = \frac{x_L}{x_L'} = \frac{x_1}{x_1'} = \frac{x_2}{x_2'} = \frac{y_2}{y_2'} = \frac{y_{2\,max}}{y_{2\,max}'} \quad ,$$

all corresponding times, by the time scale factor

$$t^* = \frac{t_1}{t_1'} = \frac{t_o}{t_o'} \quad ,$$

and all corresponding forces, by the force scale factor

$$F^* = \frac{F_L}{F_L'} = \frac{F_A}{F_A'} = \frac{F_B}{F_B'} = \frac{F_{L\,max}}{F_{L\,max}'} \quad ,$$

then the shapes, forces, and times of the model and the prototype will be similar. If the model's time axis is expanded by the time scale factor t^*, and the model's deflection axis by the length scale factor ℓ^*, then the model's deflection-time curve, $y_2'(t')$, measured at a distance x_2' , will collapse with the prototype's deflection-time curve, $y_2(t)$, measured at a distance x_2. Or, if the model's time axis is expanded by the time scale factor t^* and the model's force axis by the force scale factor F^*, then the model's force-time curve, $F_L'(t')$, measured at distance x_1' , will coincide with the prototype's force-time curve, $F_L(t)$, measured at distance x_1.

Besides lengths, forces, and times, there are other quantities to be scaled, such as speeds, accelerations, moments, stresses, and kinetic energies. We will now show that the scale factors of these "secondary" quantities are products of the scale factors of length, force, and time.

Primary and Secondary Scale Factors

In scale modeling, we are concerned only with quantities defined as products of five (or less, if less will suffice) "primary" quantities, each raised to the appropriate power. This statement in no way restricts the modeling of engineering systems; it is but the consequence of our standard system of measurement.[1] The five primary quantities are those of length, time, force,

---

[1] For a detailed discussion, see any good text on dimensional analysis.

temperature, and electric current.[1]   Therefore, only the "primary" scale
factors of these five quantities need be accounted for; all other, "secondary"
scale factors are easily derived from them.

A few examples will demonstrate the relations among the five primary and the
nearly infinite number of secondary scale factors. We stated at the beginning
that homologous behavior with respect to speed would require a constant speed
scale factor for all corresponding speeds, that is,

$$v^* = \frac{v_1}{v_1'} = \frac{v_2}{v_2'} = \frac{v_3}{v_3'} \cdots = \frac{v_n}{v_n'}$$

Speed can be expressed as the first derivative of length with respect to time,
so that

$$v^* = \frac{d\ell_1/dt_1}{d\ell_1'/dt_1'} = \frac{d\ell_2/dt_2}{d\ell_2'/dt_2'} = \cdots$$

Geometrical and temporal homology requires, however, that

$$\ell_1 = \ell^* \ell_1' , \; \ell_2 = \ell^* \ell_2' , \text{ etc., and } t_1 = t^* t_1' , \; t_2 = t^* t_2' , \text{ etc.,}$$

so that

$$v^* = \frac{\ell^*}{t^*} \frac{d\ell_1'/dt_1'}{d\ell_1'/dt_1'} = \frac{\ell^*}{t^*} \frac{d\ell_2'/dt_2'}{d\ell_2'/dt_2'} = \cdots = \frac{\ell^*}{t^*}$$

In the same way, any secondary scale factor can be derived from two or
more primary ones. One can easily show that

area scale factor $A^* = \ell^{*2}$

acceleration scale factor $a^* = \ell^*/t^{*2}$

moment scale factor $M^* = F^* \ell^*$

power scale factor $P^* = F^* v^* = F^* \ell^*/t^*$

In general, secondary scale factor $= \ell^{*n_1} t^{*n_2} F^{*n_3} \theta^{*n_4} i^{*n_5}$. Table 1

---

[1]The International Systems of Units (SI) defines six basic units: length,
time, mass, temperature, electric current, and luminous intensity. No
scale model experiments are known to the author where luminous intensity
is scaled other than in simple architectural illumination studies. Therefore, because of its very restricted use in scale modeling, luminous
intensity is disregarded here so that not more than five primary quantities
need be considered. Instead of mass, we are using here the concept of force
as a primary quantity. The SI unit of force is the newton: $1 \text{ N} = 1 \text{ kg m/s}^2$.

lists exponents n_1 through n_5 for a number of frequently used secondary scale factors.

TABLE 1. Some Secondary Scale Factors in Terms of Primary Scale Factors.

SECONDARY SCALE FACTORS	LENGTH ℓ*	TIME t*	FORCE F*	TEMPER-ATURE Θ*	ELECTRIC CURRENT i*
ANGLE, ψ*	0	0	0	0	0
STRAIN, ε*	0	0	0	0	0
AREA, A*	2	0	0	0	0
VOLUME, V*	3	0	0	0	0
MOMENT OF AREA, I_A^*	4	0	0	0	0
REVOLUTIONS PER UNIT TIME, N*; FREQUENCY, f*	0	-1	0	0	0
ANGULAR VELOCITY, $\dot{\psi}$*, ω*	0	-1	0	0	0
ANGULAR ACCELERATION, $\ddot{\psi}$*, $\dot{\omega}$*	0	-2	0	0	0
VELOCITY, v*	1	-1	0	0	0
ACCELERATION, a*	1	-2	0	0	0
RATE OF VOLUMETRIC DISCHARGE, $\dot{V}$*	3	-1	0	0	0
LINEAR IMPULSE, I*; MOMENTUM, P*	0	1	1	0	0
ANGULAR IMPULSE, I_{ang}^*; MOMENT OF MOMENTUM, H*	1	1	1	0	0
MASS, m*	-1	2	1	0	0
MASS MOMENT OF INERTIA, I_m^*	1	2	1	0	0
TORQUE, T*; MOMENT, M*	1	0	1	0	0
WORK, W*; HEAT QUANTITY, Q*; ENTHALPY, H*; ENERGY, E*	1	0	1	0	0
RATE OF HEAT FLOW, $\dot{Q}$*	1	-1	1	0	0
HEAT FLUX, $\dot{Q}$*/A*, q*	-1	-1	1	0	0
POWER, P*	1	-1	1	0	0
STRESS, σ*, τ*; PRESSURE, p*	-2	0	1	0	0
ENTROPY, S*	1	0	1	-1	0
ELECTRIC CURRENT DENSITY, j*	-2	0	0	0	1
ELECTRIC CHARGE, q*	0	1	0	0	1
ELECTRIC FIELD STRENGTH, E*	0	-1	1	0	-1
MAGNETIC FIELD STRENGTH, H*	-1	0	0	0	1
ELECTRIC DISPLACEMENT, D*	-2	1	0	0	1
MAGNETIC FLUX DENSITY, B*	-1	0	1	0	-1
EMF, VOLTAGE, POTENTIAL DIFFERENCE, V*	1	-1	1	0	-1

Applied to the previously discussed beam model, Table 1 demonstrates clearly
that from the given scale factors of length, time, and force, all the secondary scale factors (up to and including the stress scale factor) can be
easily computed.

Table 2 lists major physical fields, arranged according to the number of
participating primary scale factors. All fields listed are amenable to
scale modeling. The most extensive uses of scale modeling are in the fields
of statics, dynamics, thermodynamics, and heat and mass transfer. Models
involving electric current are less popular, and geometric and kinematic
models are trivial.

TABLE 2. Involvement of Primary Scale Factors in Various Branches of Physics.

FIELD	PRIMARY SCALE FACTOR				
	LENGTH ℓ^*	TIME t^*	FORCE F^*	TEMPER-ATURE θ^*	ELECTRIC CURRENT i^*
GEOMETRY	X	—	—	—	—
KINEMATICS	X	X	—	—	—
STATICS	X	—	X	—	—
DYNAMICS	X	X	X	—	—
THERMODYNAMICS	X	—	X	X	—
HEAT AND MASS TRANSFER	X	X	X	X	—
ELECTROSTATICS	X	—	X	—	X
ELECTRODYNAMICS ELECTROMAGNETICS	X	X	X	—	X
MAGNETOHYDRODYNAMICS	X	X	X	X	X

Representative Quantities and Pi-numbers

Table 2 demonstrates again that not more than five primary scale factors
need be considered. If these primary scale factors can be implemented, then
all secondary scale factors are automatically satisfied, too, and we have a
scale model.

These requirements look reassuringly simple. But as soon as we encounter a
practical problem, such as flow around a blunt body, we realize how difficult
modeling can become and how much we are asking: every element of the reduced
or enlarged model must repeat at every instant the flow of the prototype on
different but constant scales of length, time, and force. How can this be
done? To help answer this question, we will introduce two new concepts --
representative quantities and pi-numbers.

The earlier discussed general scaling requirement of

$$g^* = \frac{g_1}{g_1'} = \frac{g_2}{g_2'} = \cdots = \frac{g_n}{g_n'}$$

can be contracted to the shorthand notation of

$$g^* = \frac{g_r}{g_r'}$$

with g_r and g_r' representing any two corresponding quantities of the
same kind. For instance, $\ell^* = \ell_r / \ell_r'$ indicates that any two corresponding
lengths, distances, deformations, or displacements of model and prototype
must obey the same length scale factor. Likewise, $\theta^* = \theta_r / \theta_r'$ means that any
two corresponding wall temperatures, average temperatures, ambient temperatures, temperature differences, etc. must follow the same temperature scale
factor. Or $p^* = p_r / p_r'$ signifies that any two corresponding pressure drops,
initial pressures, stagnation pressures, etc. must be covered by the same
pressure scale factor.

The quantities ℓ_r and ℓ_r', θ_r and θ_r', p_r and p_r', and so forth are called
representative quantities in this book; being of major importance in scale
modeling, their application must be well understood. We will apply them
to the scale factor relations discussed in the previous section.

Any of the scale factor relations in Table 1 can be expressed in terms of
representative quantities. For instance,

$$v^* = \frac{\ell^*}{t^*}$$

can be expressed as

$$\frac{v_r}{v_r'} = \frac{\ell_r}{\ell_r'} \frac{t_r'}{t_r}$$

This relation can be easily converted into the statement of

$$\frac{v_r t_r}{\ell_r} = \frac{v_r' t_r'}{\ell_r'}$$

with all representative quantities of the prototype transferred to one side
and the corresponding representative quantities of the model to the other.
In scale modeling, dimensionless products of this kind, composed of representative quantities and required to be equal for model and prototype, play
a key role. They are therefore accorded a special name, "pi-number," and
denoted by the Greek letter π. The quantities in a pi-number need not be
labeled by the subscript "r" since their representative nature is understood.
Thus, the relation just referred to is indicated by the expression

$$\pi = \frac{v\,t}{\ell}$$

A representative quantity in a pi-number can be substituted by any like
quantity of the given phenomenon to be modeled. For instance, if we apply
the pi-number of $\pi = v t / \ell$ to the vibrating beam, discussed earlier, we can
replace the representative velocity v by the peak velocity of the imposed
external load; the representative time t , by the period of vibration; and
the representative length, by the length of the beam. The specific numerical value of a pi-number depends on the numerical values of the quantity
we choose to substitute; similarity, however, is only assured if this
specific value is maintained the same for both model and prototype.

Contemplating the steps that led us to the pi-number $\pi = v t / \ell$, we recognize
the following stations: we began with the claim that the velocities of
corresponding elements of model and prototype are governed by the velocity
scale factor. Since velocity, being the first derivative of length with
respect to time, is a secondary quantity, we were able to express the velocity
scale factor as the ratio of the length and the time scale factor: from

$v = d\ell/dt$, we derived $v^* = \ell^*/t^*$. Introducing the concept of representative quantities, we transformed the scale factor relation into the pi-number
$\pi = vt/\ell$.

In practical modeling, these steps are substantially shortened by converting
all quantities of a governing relation directly into representative quantities.
For example, in the governing relation of $v = d\ell/dt$, we convert the speed,
v, directly into a representative speed; the length differential, $d\ell$, into
a representative length; and the time differential, dt, into a representa-
tive time. Hence, $v = d\ell/dt$ is converted into $v \triangleq \ell/t$. The sign $\triangleq$ indicates that all terms are representative ones.[1] It also indicates that all
terms if assembled in product form constitute a pi-number. Thus, $\pi = vt/\ell$.

Equipped with this abbreviating mechanism, we are now ready to approach, in
the next section, an answer to the basic question asked earlier: how do we
select the correct primary scale factors so that homologous behavior of model
and prototype is assured?

The Law Approach

If all scale factors were unity, model tests -- if we could call them that --
would be simple: we would need only to repeat the experiment. That is, in
the case of flow around a blunt body, we would again position the same blunt
body at the same angle, use the same fluid, set up the same boundary and
initial conditions -- in short, duplicate all controllable quantities. The
rest we could leave to the laws of nature, which would unfailingly reproduce
the same event.

In a scale model with scale factors different from unity, the process is
essentially the same: the same laws governing the prototype must prevail
in the model, except that all model quantities must be scaled in accordance
with the primary scale factors. The question is then, how do we select the
correct primary scale factors, so that the laws governing the prototype also
govern the model?

---

[1]The symbol $\triangleq$ reads "is equivalent to."

The answer will be provided by what we here call the law approach: the
pertinent pi-numbers are directly derived from the governing laws. The law
approach is best introduced by the example of a vibrating beam.[1] We want
a model that would predict the damp-out time of the prototype's free vibrations. Since the phenomenon of the vibrating beam is one of elasticity,
inertia, and internal friction, the model must conform to three physical
laws.

(1) Elasticity is described by Hooke's law. Assuming negligible
     influence of Poisson's ratio, we relate stresses and strains
     through
     $$\sigma = E\epsilon$$

     where σ is stress, E is Young's modulus, and ϵ is strain.

(2) The inertial force on any element is ruled by Newton's law,
     $$dF = dm \cdot a$$
     where F is force, m is mass, and a is acceleration.

(3) Internal friction, as expressed by the energy loss per unit
     volume and per cycle, is (hypothetically) proportional to
     the third power of maximum stress, σ_m , regardless of
     frequency,

     $$dU = dV \cdot c \cdot \sigma_m^3$$

where dU is the energy dissipation per cycle of the volume, dV , and c is a
material constant.

In deriving pi-numbers from these three laws, we utilize the shorthand
technique demonstrated at the end of the preceding section. Accordingly,
we express the governing laws first in representative terms; then, in the
form of pi-numbers.

---

[1] We remind the reader that we do not supply rigid proofs for every assertion;
proofs are confined to "eating the pudding."

Governing law	in representative terms	as pi-number
$\sigma = E\epsilon$	$\sigma \triangleq E\epsilon$	$\pi_I = \dfrac{\sigma}{E\epsilon}$
$dF = dm \cdot a$	$F \triangleq ma$	$\pi_{II} = \dfrac{F}{ma}$
$dU = dV \cdot c \cdot \sigma_m^3$	$U \triangleq Vc\,\sigma^3$	$\pi_{III} = \dfrac{U}{Vc\,\sigma^3}$

In the given form, the three pi-numbers are hardly useful for model design; instead of in terms of stress, mass, and energy, they should be expressed in terms that are easy to work with such as length, velocity, and force.   The necessary modifications are usually accomplished with the help of representative relations among primary and secondary quantities, as indicated in the flow diagram.[1]

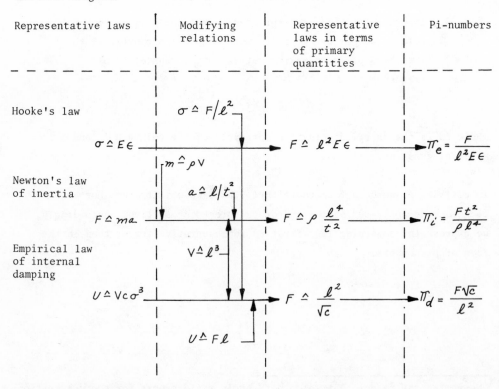

| Representative laws | Modifying relations | Representative laws in terms of primary quantities | Pi-numbers |

Hooke's law

$\sigma \triangleq F/\ell^2$

$\sigma \triangleq E\epsilon$ $\longrightarrow$ $F \triangleq \ell^2 E\epsilon$ $\longrightarrow$ $\pi_e = \dfrac{F}{\ell^2 E\epsilon}$

$m \triangleq \rho V$

Newton's law of inertia

$a \triangleq \ell/t^2$

$F \triangleq ma$ $\longrightarrow$ $F \triangleq \rho\,\dfrac{\ell^4}{t^2}$ $\longrightarrow$ $\pi_i = \dfrac{Ft^2}{\rho\ell^4}$

Empirical law of internal damping

$V \triangleq \ell^3$

$U \triangleq Vc\,\sigma^3$ $\longrightarrow$ $F \triangleq \dfrac{\ell^2}{\sqrt{c}}$ $\longrightarrow$ $\pi_d = \dfrac{F\sqrt{c}}{\ell^2}$

$U \triangleq F\ell$

[1] The author has found helpful the use of flow diagrams to clarify more complex model design problems

We now have available three independent "principal" pi-numbers[1] that if
implemented will ensure similarity in, and permit prediction of, all modeled
quantities of elasticity, inertia, and material damping. Implementation can
be accomplished in various ways. Choice of equal material for model and
prototype, for instance, requires that $E = E'$, $\rho = \rho'$, and $c = c'$, and reduces
the first pi-number or representative law to $F \triangle \ell^2 \epsilon$, or, since geometrical
similarity requires equal strains of model and prototype, to $F \triangle \ell^2$. This
relation satisfies the third pi-number, π_d (since $c = c'$), and reduces the
second pi-number, π_i , to $t \triangle \ell$. Hence, with the representative time
expressed as damp-out time, t_o , we have $t_o / t_o' = \ell^*$; i.e., the damp-out time
of the model will be ℓ^* - times shorter than that of the prototype if the
same materials are used for both.

In practical modeling, pi-numbers directly derived from physical laws are
usually modified by eliminating a representative quantity common to all
pi-numbers. By eliminating the representative force, for instance, we obtain
two pi-numbers from the three governing laws, as shown in the following flow
diagram.

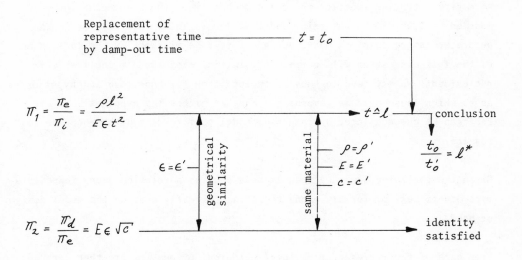

---

[1] In this book, pi-numbers derived directly from physical laws are called
principal pi-numbers; see next section.

With the same-material stipulation, we again arrive at the conclusion that
the prototype's damp-out time can be predicted by multiplying the model's
measured damp-out time by the length scale factor. Of course, the total
number of principal pi-numbers necessary for modeling all aspects of the
vibrating beam remains three even if eliminating the representative force
(or any other representative quantity common to all three pi-numbers) yields
only two principal pi-numbers. A third, independent pi-number is still present,
in the form of Π_e , Π_i , or Π_d . From Π_1 and Π_2 , the time scale factor is
determined; from Π_e or Π_i or Π_d , the force scale factor.

Note that independent pi-numbers of the same phenomenon can be multiplied
or divided with each other or raised to a power without losing their
independence. The number of independent principal pi-numbers, however,
always remains the same (e.g., three in the vibrating beam, example) no mat-
ter how the initially derived pi-numbers are combined.

Guided by the discussed example, we are now ready to outline the general
approach applied throughout this book, an approach we call the "law
approach" because it is founded directly on physical laws.

As a first step, we identify the basic mechanism of the phenomenon to be
modeled by listing the governing physical laws. These laws must be known
before we can contemplate model design, for if we misidentify the laws, none
of the following steps will correct our initial mistake. If the laws are
not evident, we may have to substitute intuition for knowledge and hypothesize
as to which laws would be governing. Whether or not the hypothesis is correct can be checked later by comparing model test results with prototype test
results.

Given the governing laws, we then express each as a relation among representative quantities, preferably among those that directly bear on the model design
at hand.

From each of the representative laws, principal pi-numbers are then derived.
Frequently, these pi-numbers are combined into other pi-numbers by

multiplication or division. For instance, it is common practice to eliminate
a representative quantity that occurs in most or all pi-numbers derived from
laws, and thus obtain $(n-1)$ pi-numbers from n original ones. The preferred
candidate for elimination is the representative force because most physical
laws used in engineering can be traced to the concept of force in form of
pressure, torque, impulse, mechanical energy, and power. Another favored
candidate is the representative energy, mostly heat energy. Regardless of
how original pi-numbers are combined into new ones, however, their total
number must equal the number derived originally from the governing laws.

Finally, we impress all principal pi-numbers onto the model, so that we
obtain a scale model. From our observations of the scale model we can then
predict prototype performance.

Principal and Common Pi-numbers

A pi-number is a dimensionless product of representative quantities, as
pointed out. There are as many pi-numbers as relations existing among representative quantities. All relations among primary and secondary scale factors,
for instance, can be expressed as pi-numbers. Thus, the scale factor relation
among power, force, length, and time (Table 1), $P^* = \ell^* F^* / t^*$, expressed in
representative quantities, $P \triangleq \ell F / t$, leads to the pi-number, $\pi = Pt/(\ell F)$.
These and similar pi-numbers not associated with physical laws need not be
spelled out explicitly; they are taken for granted. We will call them
common pi-numbers. The only pi-numbers that cannot be taken for granted
but must be found anew for each problem are those derived from the
governing physical laws. In this book, pi-numbers derived from laws are
called *principal* pi-numbers to indicate their key importance in scale
modeling. In recognition of the importance of principal pi-numbers, most
of them have been given the names of eminent scientists and engineers. In
Appendix B at the end of this book, sixty named principal pi-numbers
are catalogued, together with their major fields of application, their
constituting physical laws, and their basic definitions.

As an illustration of named principal pi-numbers, we will discuss the
Reynolds number, the Cauchy number, and the Fourier number. Each is composed from two physical laws.[1]

The *Reynolds number* is composed from Newton's law of inertia and the law
for shear stress of a Newtonian fluid. Newton's law of inertia and its
representative form have been discussed earlier: from the law, $F = ma$,
where m is mass and a is acceleration, we deduced the representative form
of $F \triangleq ma$ which, with the help of the representative relations of $a \triangleq l/t^2$,
$m \triangleq \rho l^3$, and $v \triangleq l/t$ can be transformed into $F \triangleq \rho l^2 v^2$. The same result
can be obtained from any other form of Newton's law; for instance from the
impulse equation of a particle. The impulse I of a force F over a time
interval $(t_2 - t_1)$ equals the change of momentum mv during the time
interval,

$$I = \int_{t_1}^{t_2} F \, dt = m(v_2 - v_1) \, .$$

The representative version of this expression is $Ft \triangleq mv$, which is identical with $F \triangleq ma$, and, hence, with $F \triangleq \rho l^2 v^2$. We could also start from the
change of kinetic energy, E , of a particle moving from position r_1 to r_2 ,

$$E = \int_{r_1}^{r_2} F \, dr = \frac{m}{2}(v_2^2 - v_1^2) \, .$$

Expressed in representative terms, this relation transforms into

$$E \triangleq Fl \triangleq mv^2$$

Hence, we have, with $l \triangleq v \, t$ and $a \triangleq v/t$, again $F \triangleq ma$, which leads to
$F \triangleq \rho l^2 v^2$, as shown. From this representative version, a principal pi-
number, the so-called Newton number, can be derived

$$\underline{Ne} = \frac{F}{\rho l^2 v^2}$$

---

[1] Principal pi-numbers can be composed from any number of laws, as will
be seen.

To arrive at the Reynolds number, the law for shear stress of a Newtonian fluid must be converted into a representative relation. The law can be expressed as

$$\tau_{xy} = \mu \left(\frac{\partial v_x}{\partial y} + \frac{\partial v_y}{\partial x} \right)$$

with two more equations obtained by cyclic permutation of the subscripts, where μ is the coefficient of viscosity, and τ is the shear stress. We find by inspection that all terms can be transformed into the representative relation of $\tau \triangleq \mu \frac{v}{\ell}$. Since $\tau \triangleq F/\ell^2$, we have $F \triangleq \mu \ell v$. The resulting principal pi-number, $\pi_v = \frac{F}{\mu \ell v}$, has no name. However, by dividing this pi-number by the Newton number, that is, by eliminating the representative force, we obtain the Reynolds number

$$\underline{Re} = \frac{\pi_v}{Ne} = \frac{\rho \ell v}{\mu}$$

The Reynolds number is often interpreted as the ratio of representative inertial force and representative viscous force. This statement is only correct if "representative inertial force" is understood as a short expression for the representative product of $\rho \ell^2 v^2$; and "representative viscous force," as a short expression for the representative product of $\mu \ell v$. Then, instead of the two representative laws, $F \triangleq \rho \ell^2 v^2$, and $F \triangleq \mu \ell v$, one can simply write, $F \triangleq F_i$ and $F \triangleq F_v$, with $F_i \equiv \rho \ell^2 v^2$ and $F_v \equiv \mu \ell v$. With these definitions, $\underline{Re} = F_i/F_v$. We will not use this notation here, however.

All these derivations can be expressed succinctly in a flow diagram.

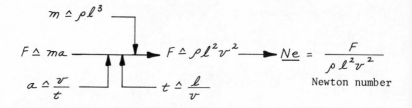

Newton's law of inertia

$$F \triangleq ma \longrightarrow F \triangleq \rho \ell^2 v^2 \longrightarrow \underline{Ne} = \frac{F}{\rho \ell^2 v^2}$$

$$a \triangleq \frac{v}{t} \qquad t \triangleq \frac{\ell}{v}$$

Newton number

Shear stress of Newtonian fluid

$$\tau \triangleq \mu \frac{v}{\ell} \longrightarrow F \triangleq \mu \ell v \longrightarrow \pi_v = \frac{F}{\mu \ell v}$$

$$\tau \triangleq \frac{F}{\ell^2}$$

Reynolds number

$$\underline{Re} = \frac{\pi_v}{\underline{Ne}} = \frac{\ell v}{\nu}$$

where $\nu \equiv \dfrac{\mu}{\rho}$ is the kinematic viscosity.

A second illustration of a principal pi-number is the derivation of the *Cauchy number*, which is constituted from Newton's law of inertia and Hooke's law of elasticity of isotropic solids. This law can be expressed as

$$\epsilon_x = \frac{1}{E} \left[\sigma_x - \nu \left(\sigma_y + \sigma_z \right) \right]$$

with two more equations obtained by cyclic permutation of the subscripts, where ϵ is strain, σ is stress, E is Young's modulus, and ν is Poisson's ratio. Inspecting this law we note that it yields two representative relations,

$$\epsilon \triangleq \frac{\sigma}{E} \qquad \text{and} \qquad \epsilon \triangleq \nu \frac{\sigma}{E} \quad .$$

Since it is impossible to satisfy both relations (unless $\nu = \nu'$), the influence of Poisson's ratio on the response of elastic structures is usually neglected.[1] Then, only the first version remains. With this version and

---

[1] Another possibility is offered by the "same material" approach, see Chapter 3.

the representative version of Newton's law derived earlier, the Cauchy
number can be developed in a flow diagram.

$$\epsilon \triangleq v \frac{\sigma}{E}$$ Influence of v neglected

$$\sigma \triangleq \frac{F}{\ell^2}$$

Hooke's law $\epsilon \triangleq \frac{\sigma}{E}$ $\longrightarrow F \triangleq \ell^2 E \longrightarrow \pi_e = \frac{F}{\ell^2 E}$

Geometric
similarity $\epsilon = \epsilon'$

Newton's law
of inertia $F \triangleq \rho \ell^2 v^2 \longrightarrow Ne = \frac{F}{\rho \ell^2 v^2}$

Newton number

Cauchy number $Ca = \dfrac{\pi_e}{Ne} = \dfrac{\rho v^2}{E} = \left(\dfrac{v}{a_s}\right)^2$

with $a_s = \sqrt{E/\rho}$, the speed of sound in the elastic material.

A last example of a principal pi-number is the derivation of the *Fourier
number* from two physical laws: Fourier's law of heat conduction, and the
law of heat capacity. Fourier's law is usually expressed as

$$q = -k \; grad \; \theta$$

where q is the heat flux (heat energy per unit area and unit time), k
is the thermal conductivity, and θ is temperature. Written in representative terms, Fourier's law takes the form of

$$q \triangleq k \frac{\theta}{\ell} \; .$$

The law of heat capacity can be expressed as

$$Q = c_p \; m \Delta \theta$$

where Q is heat energy, c_p is the specific heat at constant pressure, m
is mass, and $\Delta \theta$ is the temperature difference. In representative terms,

$$Q \triangleq c_p \; m \theta$$

The derivation of the Fourier number from these two representative laws follows the pattern established in the previous examples.

$$q \triangleq \frac{Q}{l^2 t}$$

Fourier's law of heat conduction

$$q \triangleq k \frac{\theta}{l} \qquad\qquad Q \triangleq \frac{k l^2 \theta}{v} \qquad\qquad \pi_k = \frac{Q v}{k l^2 \theta}$$

$$t \triangleq \frac{l}{v}$$

$$m \triangleq \rho l^3$$

Law of heat capacity

$$Q \triangleq c_p \, m \theta \qquad\qquad Q \triangleq c_p \rho \, l^3 \theta \qquad\qquad \pi_c = \frac{Q}{c_p \rho \, l^3 \theta}$$

Fourier number

$$F_o = \frac{\pi_c}{\pi_k} = \frac{\alpha}{l v}$$

where $\alpha \equiv \dfrac{k}{c_p \rho}$ is defined as thermal diffusivity.

From these examples (and from the many principal pi-numbers developed in the Appendix), we conclude that each physical law delivers one or more pi-numbers in the form of dimensionless "power groups" constituted from five (or less) primary representative quantities.

$$\pi = c \, l^{n_1} t^{n_2} F^{n_3} \theta^{n_4} i^{n_5}$$

where the exponents n_1 through n_5 are typical for the given law, and the factor C is a constant or a material property.

For instance, in our previous examples, we found the following principal pi-numbers:

for Newton's law of inertia ,

$$Ne = \frac{F}{\rho l^2 v^2} \quad;$$

for the law of shear stress of a Newtonian fluid ,

$$\pi_V = \frac{F}{\mu l v} \quad;$$

for Hooke's law of isotropic solids,

$$\pi_{e1} = \frac{F}{E\ell^2} \quad and \quad \pi_{e2} = \frac{F\nu}{E\ell^2}$$

(note that Hooke's law is an example of a law that evolves more than one principal pi-number);

for Fourier's law of heat conduction,

$$\pi_k = \frac{Q\nu}{k\ell^2\theta} \quad ;$$

for the law of heat capacity,

$$\pi_c = \frac{Q}{c_p\rho\ell^3\theta} \quad .$$

If reduced to relations among primary representative quantities, each of the quoted principal pi-numbers can be expressed in terms of a dimensionless power group. For instance, for the Newton number: $c = \rho^{-1}$, and $n_1 = -4$, $n_2 = 2$, $n_3 = 1$, $n_4 = n_5 = 0$. Or for π_c : $c = (c_p\rho)^{-1}$, $n_1 = -2$, $n_2 = 0$, $n_3 = 1$, $n_4 = -1$, $n_5 = 0$.

Dimensionless power groups and pi-numbers are usually developed by applying dimensional analysis -- a branch of applied mathematics with a well-developed theoretical background (including the Pi-Theorem). These theories need not concern us here; the law approach will fully provide us with all the information that is necessary to design and use scale models.[1] A few remarks on dimensional analysis are made in the next section.

The Equation and the Parameter Approaches

Scale model experiments must be designed such that model and prototype follow the same physical laws, as pointed out. No more understanding of a phenomenon is required than to know the physical laws that govern the behavior of each of its elements, and from these laws derive the principal pi-numbers. If the principal pi-numbers are derived directly from

---

[1]Texts on dimensional analysis are referenced at the end of Part I.

the governing laws, we call this method the law approach. The two other
more commonly used approaches, the equation approach and the parameter
approach, are based on the governing laws also, but in less direct ways.

We can demonstrate the three approaches with the following problem.[1] If a
six-pound roast of beef requires three hours of cooking time,[2] how many
hours would a three-pound roast require at the same temperature? We assume
geometrical similarity between the six-pound "prototype" and the three-
pound "model."

First, in the law approach, two laws are involved -- Fourier's law of heat
conduction, and the law of heat capacity. Both laws have previously been
expressed as principal pi-numbers, Fourier's law as

$$\pi_k = \frac{Q v}{k \ell^2 \theta}$$

and the law of heat capacity as

$$\pi_c = \frac{Q}{c \rho \ell^3 \theta}$$

where Q is heat energy, v is velocity, k is thermal conductivity, ℓ is
length, θ is temperature, c is specific heat, ρ is density; all quantities
are representative.

By eliminating the representative heat, Q , we arrive at the Fourier number
and, from it, at the time needed to cook the small roast.

---

[1] This problem has been adapted from S.J. Kline, *Similitude and Approximation Theory*, McGraw-Hill, New York, 1965, p. 25.

[2] Cooking time may be defined as the total time necessary to raise the temperature of every particle of the roast to a temperature close to the oven temperature and then hold it for a specified amount of time.

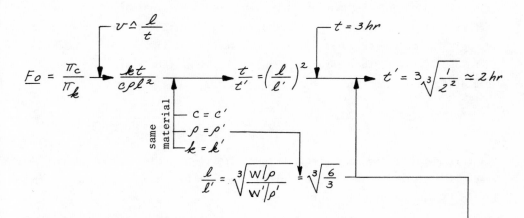

where W is the roast's weight. The heat needed to cook the model roast
follows from π_c or π_k.

$$\pi_c = \frac{Q}{c\rho\ell^3\theta} \qquad\qquad\longrightarrow\qquad Q' = Q\left(\frac{\ell'}{\ell}\right)^3 = \frac{Q}{2}$$

$$c = c'$$
$$\rho = \rho'$$

same mat.

$$\theta = \theta'$$

same temp.

Hence, the heat is simply proportional to the weight of the roast.

The equation approach starts with governing equations. The process of
formulating the equations requires not only having an intimate knowledge of
the governing laws, but also knowing fairly well how they interact with each
other. We usually apply the governing laws to a representative small element, so that we can express the interrelations among the laws in form of
differential equations. Here, the governing equation is the well-known
differential equation for heat conduction, derived from Fourier's law and
the law of heat capacity.

$$\frac{\rho c}{k}\frac{\partial\theta}{\partial t} = \frac{\partial^2\theta}{\partial x^2} + \frac{\partial^2\theta}{\partial y^2} + \frac{\partial^2\theta}{\partial z^2}$$

ρ, c, and k are assumed to be independent of temperature.

From this equation, pi-numbers can be extracted by a transformation process.
Different transformation methods are advocated by different researchers; here
we apply the method of scale factors. We state that if the model phenomenon
is to be similar to the prototype phenomenon, it must be governed by the
same differential equation,

$$\frac{\rho'c'}{k'} \frac{\partial \theta'}{\partial t'} = \frac{\partial^2 \theta'}{\partial x'^2} + \frac{\partial^2 \theta'}{\partial y'^2} + \frac{\partial^2 \theta'}{\partial z'^2}$$

All corresponding lengths, times, temperatures, etc., are assumed to have
respective scale factors. Substituting these scale factors into the governing model equation yields

$$\frac{\rho c}{k} \frac{k^*}{\rho^* c^*} \frac{\partial \theta}{\partial t} \frac{t^*}{\theta^*} = \left(\frac{\partial^2 \theta}{\partial x^2} + \frac{\partial^2 \theta}{\partial y^2} + \frac{\partial^2 \theta}{\partial z^2} \right) \frac{\ell^{*2}}{\theta^*}$$

Comparing this equation with the prototype equation yields

$$\ell^{*2} = \frac{k^*}{\rho^* c^*} t^*$$

or

$$\frac{kt}{\rho c \ell^2} = \frac{k't'}{\rho'c'\ell'^2} \quad ; \quad i.e. \quad \pi = \frac{kt}{\rho c \ell^2}$$

The result is the one of the two pi-numbers we derived from the law approach,
i.e., the Fourier number.

Last, in the popular parameter approach, which, in a way, is the most difficult of the three, we start with a list of the important parameters.[1] The
basis of the parameter approach is that the governing equations are dimen-
sionally homogeneous, and that, because of this, the governing parameters
constituting the (unknown) equation can be arranged in dimensionless pi-
numbers. That this arrangement is possible is asserted by Buckingham's
Pi-Theorem. The two methods most widely practiced are those of Lord Rayleigh's

---

[1]The term "parameter," as traditionally used in the parameter approach,
has the same application as the term "representative quantity" used in
the law approach.

and E. Buckingham's.[1] Here, we will not go into a discussion of both methods,
but we must emphasize, that no method of dimensional analysis can produce
new arguments after the parameters that allegedly characterize the phenomenon
have once been chosen. If we should pick erroneous parameters or forget an
important one, no subsequent analysis of the parameters' dimensions can
correct our error. The real weakness of the parameter approach is that it
offers so little help in selecting the important parameters.

In our example, five significant parameters[2] for the heat transfer exist:
the time, t, required to cook the roast; characteristic dimension, ℓ, of the
roast; density, ρ; specific heat, c; and thermal conductivity, k. Now, the
four primary parameters[3] that describe this particular phenomenon are
length, ℓ, time, t, heat energy, Q, and temperature, θ. Although temperature
is not listed as a parameter, it is inherent in the dimensions of both thermal
conductivity and specific heat.

In the routine of obtaining pi-numbers from the governing parameters, we
first display the dimensions of the parameters in a matrix. Next, we check
the "rank" of the dimensional matrix by picking a determinant of the matrix
and computing its numerical value. If it is nonzero, and if all determinants
of an order larger than that of the picked determinant are zero, then the
rank of the matrix is equal to the order of the nonzero determinant, shown
below to be 4.

---

[1] In 1915, Lord Rayleigh ("The principle of similitude," *Nature*, 95, 2368
[Mar 1915] 66-68; and 95, 2389 [Aug 1915] 644) laid the foundations of
dimensional analysis followed soon after by E. Buckingham ("The principle
of similitude," *Nature*, 96, 2406 [Dec 1915] 396-397), who stated his pi-
theorem. Later, P.W. Bridgman (*Dimensional Analysis*, Yale University
Press, 1922) consolidated and published the methods brought forward by
these two pioneers. Since then, dimensional analysis has been firmly
established as the foundation of model theory.

[2] Actually, six. But S.J. Kline notes that if temperature is included, the
parameter approach will not work. He states, "On the basis of this
method [parameter approach] alone, omission of temperature is a questionable
procedure. The author must admit that it has been omitted primarily because
one thus obtains a correct answer, and the justification can be provided
from more complete analysis." Thus, the parameter approach sometimes needs
to be checked by other methods (a procedure that is unnecessary if the law
approach is used).

[3] For this particular problem, energy rather than force is considered a
primary quantity.

	Parameters				
	ℓ	t	k	c	ρ
ℓ	1	0	-1	2	-5
t	0	1	-1	-2	2
Q	0	0	1	0	1
θ	0	0	-1	-1	0

Primary Parameters

$$\begin{vmatrix} 0 & -1 & 2 & -5 \\ 1 & -1 & -2 & 2 \\ 0 & 1 & 0 & 1 \\ 0 & -1 & -1 & 0 \end{vmatrix} = -2 \neq 0$$

Then, according to the pi-theorem, we expect a single dimensionless product, i.e., a single pi-number, because 5 - 4 = 1 where five is the number of parameters and four the rank of the nonzero determinant. This pi-number is a product of the listed parameters; consequently, we can state in general notation that

$$\pi = \ell^{K_1} t^{K_2} k^{K_3} c^{K_4} \rho^{K_5}$$

$$= \ell^{K_1} t^{K_2} (\ell^{-1} t^{-1} Q \theta^{-1})^{K_3} (\ell^2 t^{-2} \theta^{-1})^{K_4} (\ell^{-5} t^2 Q)^{K_5}$$

$$= \ell^{(K_1 - K_3 + 2K_4 - 5K_5)} t^{(K_2 - K_3 - 2K_4 + 2K_5)} Q^{(K_3 + K_5)} \theta^{(-K_3 - K_4)}$$

Since π is required to be dimensionless, the exponents of t, ℓ, Q, and θ must all be zero.

$$K_1 - K_3 + 2K_4 - 5K_5 = 0$$

$$K_2 - K_3 - 2K_4 + 2K_5 = 0$$

$$K_3 + K_5 = 0$$

$$K_3 + K_4 = 0$$

Since we have five unknowns but only four equations, the system of equations is under-determined. We now assume that one of the five unknown exponents, K_1 through K_5, equals 1. If we take $K_5 = 1$, then the solutions are

$$K_1 = 2, \quad K_2 = -1, \quad K_3 = -1, \quad K_4 = 1, \quad K_5 = 1$$

and

$$\pi = \frac{\ell^2 c \rho}{t k} \qquad \text{or} \qquad \frac{1}{\pi} = \frac{t k}{c \rho \ell^2}$$

Hence, we have again one of the two pi-numbers obtained by using the law approach.

With this simple heat-transfer problem, both the equation approach and the parameter approach are shown to be, in essence, indirect law approaches. The law approach is the method that uses the governing laws directly; it transforms the governing laws into relations among representative quantities, and from these relations the pi-numbers are derived. The equation approach starts from governing equations (i.e., from equations derived from the governing laws) and extracts pi-numbers by a normalizing process. The parameter approach enumerates all those parameters that compose the governing laws and subjects them to dimensional analysis; application of the pi-theorem yields again the same pi-numbers. Figure 18 diagrams the three approaches.

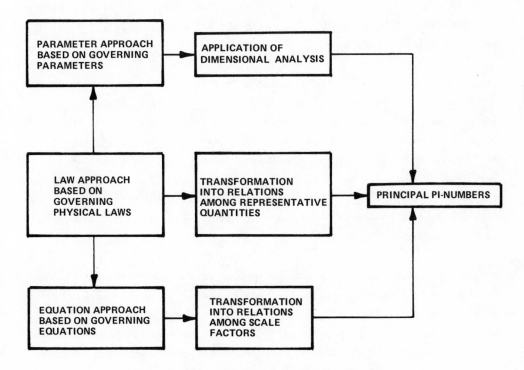

Fig. 18.   Three approaches to obtaining pi-numbers.

In this book, we emphasize the law approach because it permits the direct
identification of the governing parameters and their natural arrangement
in dimensionless power groups; i.e., in principal pi-numbers. Note also
in our example that both the equation and the parameter approach yield only
one principal pi-number, whereas two are required to describe the model
in full, a number almost automatically produced by the law approach simply
by listing the two governing laws in terms of representative quantities.

Pilot and Production Experiments

Physical laws and their interactions are established through the planned
observation of nature -- that is, through experiment. To be most effective,
experimentation should begin with a hypothesis as to which laws might be
the ones that govern the prototype being studied. Experiments will confirm
the hypothesis or reject it. Unfortunately, we must sometimes begin with
almost no idea of the nature of a phenomenon. Then, the only way to form
any opinion of its nature is to perform some "let-us-try-and-see-what-happens"
experiments, called *pre-hypothesis* experiments because they are not based upon
an intelligent hypothesis.

The pre-hypothesis experiment is certainly not amenable to scale model experiments, since it takes place before the governing laws have been determined.
And yet, many model experiments are carried out with the hope that they will
circumvent the need to establish -- or hypothesize -- the laws governing the
new phenomenon. These attempts (indirectly promoted in the parameter
approach) are futile. No scale model experiments should be designed without
a hypothesis about its basic mechanism, i.e., about its governing laws. Some
laws are easy to recognize, others evolve after study, and still others must
be conjectured.

Once the governing laws and associated principal pi-numbers are established,
we must verify them by tests. Called *scale model pilot experiments*, these
verification tests are a most painstaking but interesting part of model
studies: results of pilot experiments are compared with the results of prototype experiments to see whether they are in accordance with the hypothesized pi-numbers. If they are not, then either the original interpretation

of the phenomenon was wrong, meaning that some of the important governing laws were overlooked, or else the pilot experiments were performed improperly. Although prototype results are not always available for comparison, the hypothesized assembly of governing laws should be checked later, when the prototype is eventually built. Even wrong results, if analyzed with diligence and intuition, can be of help.

As an example of the entire process, consider the vibrations of a suspension bridge observed under the action of wind. We may hypothesize that the vibrations are caused by a dynamic interplay among its inertial forces, elastic forces, and wind pressure, and that viscous forces of the wind acting on the bridge can be neglected. We then construct a bridge model in accordance with this hypothesis and perform pilot experiments. If our hypothesis is correct, the behavior of the model will agree with the behavior of the original bridge. If it does not, then possibly viscous forces play a larger role than we expected. Now, new hypotheses must be formulated and tested until a good correlation between the behavior of the model and the full-scale bridge is obtained. Only then can we consider our hypothesis to be good, and only then can we construct a model bridge for *scale model production experiments*, confident that it will be a valid substitute for the prototype.

Thus, there are two objectives in scale model experimenting: one is to verify a hypothesis about governing laws, achieved by pilot experiments; the other is to predict the performance of a prototype, achieved by production experiments. Both are discussed in Part II. In practice, pilot and production experiments cannot be clearly separated. In every production experiment, some of the ground we tread is unknown, for no physical phenomena are exactly alike. And in each pilot experiment, with the premise that no physical phenomena are completely divorced from the rest of the known world, some of our information must be confirmed information. Nevertheless, procedures for the two kinds of experiments differ, for production experiments are based on firm information about the governing laws. The test engineer is not concerned about the significance of this or that pi-number -- all important pi-numbers are known. He is concerned with setting up test facilities, developing instrumentation, and designing test schedules.

On the other hand, pilot experiments are intended to affirm or reject
assumptions about just which physical laws do govern a phenomenon. Pilot
experiments tax our faculties: literature is to be reviewed, previous experience to be expanded, a hypothesis about the governing physical mechanism
to be formulated, crucial tests to be devised. Pilot experiments are an
exploration into the unknown that allows us, finally, to penetrate the
narrowness of an isolated process and to work near the general nature of
a phenomenon. Without understanding its basic nature, we cannot hope to
obtain good agreement between model and prototype results.

The results of all pilot experiments should eventually be checked with prototype tests, as noted. In many cases, however, a prototype is unavailable.
One example is the development of a supersonic transport that is still on
the drawing board at the time of model testing. Another is the study of
human shock tolerances performed on primate "models" because the tests are
too severe for humans. Still another is the study of spacecraft whose
environment cannot be completely duplicated on earth. In cases like these,
efforts are often made to approximate prototype tests somehow. One valid
method is to repeat pilot experiments, using various scales; if test results
of different scales agree, the possibility of their being valid substitutes of
the prototype is good. This technique, of excellent help in the development
of aircraft, is of little help in the study of human shock tolerance because
human models cannot be built to various scales. Instead, other sources of
information, such as medical reports of human accidents, must be found. Also,
mathematical analysis may be added, if possible.

Presentation of Test Results

The presentation of model test results as an empirical function among pi-
numbers, in the general form of $\pi_o = f(\pi_1, \pi_2, \ldots \pi_n)$, is common practice,
with the pi-numbers treated as variable entities. At least one of them is
always a principal pi-number reflecting the law or laws governing the
modeled system; and most likely, though not necessarily, one of them is a
common pi-number indicating a quantity of interest. Also, one of the pi-
numbers is treated as the independent variable; the rest, as dependent
variables. In the vibrating beam problem (see Section "The Law Approach"),

the common pi-number y/x_L (Fig. 17) is considered the dependent variable; the two principal pi-numbers, the independent variables.

Presenting the results of scale model tests in terms of pi-numbers has the great advantage of reducing the number of variables considerably, in our example from six (E, ρ, c, y, v, x_L) to three $(y/x_L, (\rho/E)v^2, E\sqrt{c})$, so that one family of curves suffices to predict the amplitude y at any given size of the system (represented by x_L) for any given material (represented by E, ρ, and c) at any given rate of deformation (represented by v). This reduction of the number of variables is a welcome help in scale model work, for the interrelations among the many variables (here, six) can be condensed into a few plots (here, one). However, this advantage pertains only to truly homologous systems in which all important variables are scaled in accordance with the ruling principal pi-numbers; it should not be applied to non-homologous systems. Unfortuantely, this is sometimes done.

Plotting experimental data in terms of dimensionless variables is, of course, a well-accepted practice in engineering. There are numerous dimensionless engineering terms such as efficiency, slip, and normalized force, which when applied to non-similar systems have quite useful functions; they are a convenient help in comparing data obtained from systems of the same type. However, these dimensionless ratios are always common pi-numbers whose use is not restricted to homologously scaled systems. The application of principal pi-numbers, however, must not be extended beyond truly scaled systems; the result would be confusion.

CHAPTER 3

RELAXATION OF DESIGN REQUIREMENTS

<u>Why Relaxation?</u>

If more than one principal pi-number is involved in a scaling problem, correct scaling may be difficult. From a strictly mathematical point of view, the number of principal pi-numbers identified in a given problem does not in any way restrict the free choice of primary scale factors because each additional principal pi-number introduces at least one additional physical property whose value could, in principle, be adapted to the selected primary scale factors. For instance, in the earlier example of the vibrating beam, the three principal pi-numbers developed from the three laws of inertia, elasticity, and internal damping would yield the following three scale factor relations

$$\pi_e = \frac{F}{\ell^2 E \epsilon} \quad, \quad \text{leading to} \quad E^* = \frac{F^*}{\ell^{*2}}$$

$$\pi_i = \frac{F t^2}{\rho \ell^4} \quad, \quad \text{leading to} \quad \rho^* = \frac{F^* t^{*2}}{\ell^{*4}}$$

$$\pi_d = \frac{F \sqrt{c}}{\ell^2} \quad, \quad \text{leading to} \quad c^* = \frac{\ell^{*4}}{F^{*2}}$$

The primary scale factors of F^*, ℓ^*, and t^* can be chosen arbitrarily, from a mathematical point of view; we would only have to adapt the three material properties to the three given scale factor products.

Since materials whose properties would meet these three requirements are not easy to find, however, and since the art of tailoring material properties to desired specifications is still in its infancy, we are forced to adapt the primary scale factors to the material constants available. For instance, if we decide to use the same material for both model and prototype; i.e.,

59

$E^* = \rho^* = c^* = 1$, the scale factors become $\ell^* = t^*$ and $F^* = \ell^{*2}$, as already demonstrated.

If Newton's law of gravity is added to the first three laws, yet another principal pi-number must be satisfied: from $F \triangleq \rho g \ell^3$ follows

$$\pi_g = \frac{F}{\rho g \ell^3} \quad , \quad \text{leading to} \quad \rho^* = \frac{F^*}{g^* \ell^{*3}}$$

With $g^* = 1$, the four scale factor relations could be satisfied simultaneously only if $E^* = \rho^* \ell^* = 1/\sqrt{c^*}$ and $\ell^* = t^{*2}$. But this is possible (when using the same material) only if $\ell^* = t^* = 1$, a condition that denies scaling altogether. Even with dissimilar materials, the required conditions for E^*, ρ^*, and c^* are very difficult to achieve. To avoid scaling difficulties of this sort, we try to keep the identified laws to a minimum. But since we are bound to appreciate and scale all laws governing a phenomenon, we may still be confronted with scale factor conflicts.

These considerations lead us to an important point in scale modeling: the more physical laws are involved in a phenomenon, the more principal pi-numbers are to be satisfied (i.e., the more interrelations to be reckoned with among the primary scale factors) and the less freedom exists in choosing the materials and the scale factors.

Fortunately, conflicting claims on scale factors can frequently be resolved by what we call *relaxation* ,[1] a single word meaning the resolution of scaling conflicts by using all prior knowledge of the phenomenon to be scaled. Relaxations might consist, for instance, of studiously neglecting less important laws, or of infusing the experiment with analytical knowledge, or even of dividing the whole phenomenon into smaller, manageable parts.

All scale models are relaxed models in a strict sense because we invariably disregard minute details to concentrate on essentials. Relaxations can be

---

[1] We prefer the term "relaxation," rather than "distortion," as often used. According to Webster, distortion suggests twisting out of a natural, normal, or original shape; whereas relaxation denotes a lessening of rigor, which is what we are striving for. Furthermore, distortion refers generally to geometrical shape, while relaxation can be applied to any aspect of model design.

very difficult; but without them scale model experiments would be rather
limited, as evidenced by many of the case studies in Part II. The consummate
application of relaxations makes scale model experiments possible.

Relaxation is an art rather than a formal process. Each problem must be
diagnosed separately. There are no recipes -- only some general rules. In
the course of scale modeling history, a number of relaxation methods have
evolved, not systematically but in rather diverse ways geared to specific
problems. In this chapter an attempt is made to set down a few principles
that appear to apply to most relaxation cases in one way or another. These
principles, summarized in Fig. 19, are discussed in some detail.

Identifying Essential Laws

In a scientific or engineering problem we begin with a selected and simplified part of nature, eliminating most of the marginal and nonessential laws
to arrive at a useful approximation (thus, even the full-scale test is often
a relaxed version of the true event). But even then, we may find that we
have too many laws, and that scaling is difficult or impossible. When this
happens, further relaxations are necessary.

As a first step in relaxation, it is helpful to determine whether the laws
causing scaling conflicts are governing the given system with equal or
unequal strength. If they are governing it with unequal strength, the
weakest laws can perhaps be disregarded within segments of the investigation
if not throughout its entire range. If their influences are equally strong,
none can be neglected outright, but correct results may still be attained
by skillful circumvention of the most disturbing law.

When a law's effect on total performance is uncertain, tests with a pilot
model may supply missing information. Suppose that three laws are known
to govern a phenomenon; of the three, two are strong, but the third is
alleged to be weak. Three principal pi-numbers can be formulated, with one
representing the weak law. In pilot model tests, the "weak" pi-number will
be revealed as weak (as supposed) or strong by checking its effects on the
similarity between model and prototype results. Figure 20 shows results of a

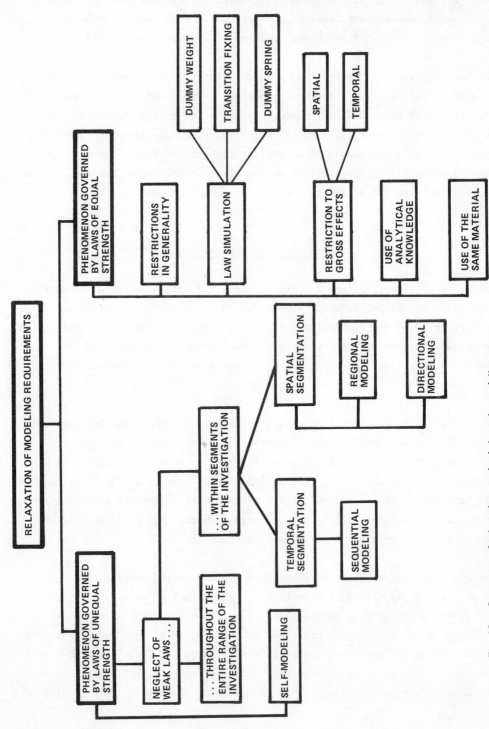

Fig. 19.   Summary of relaxation methods in scale modeling.

hypothetical test to check the effect of π_1 on the phenomenon by plotting π_3 versus π_1 , with constant π_2 . If the tests produce curve (1), we conclude that π_1 exerts a strong influence; if they produce curve (2), a weak influence; if they produce curve (3), we conclude that π_1 exerts a strong influence at low π_1 -values but a weak influence at high π_1 -values.

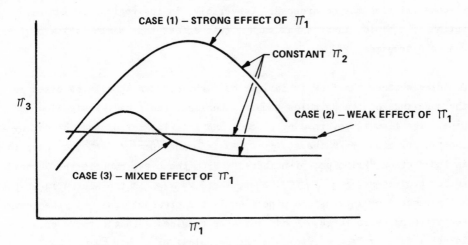

Fig. 20.   Three hypothetical effects of π_1 on π_3 .

Occasionally, the term *scale effect* is used in the literature. As G. Murphy[1] has correctly pointed out, this term does not imply that models are inherently incapable of accurately predicting prototype performance. True models reproducing *all* governing laws will always give accurate predictions. Only if scaling requirements are violated -- for instance by neglecting weak governing laws -- would the prediction accuracy suffer. Even then, the lost accuracy may be restored by applying some of the remedies described in the following sections.

Disregarding Weak Laws

Disregarding weak governing laws is perhaps the most common relaxation practice. A well-documented example is the steady-state flow of gases or fluids.

---

[1]G. Murphy, *Similitude in Engineering*, Ronald Press, New York, 1950

The flow is governed by two laws: Newton's law of inertia and the law of viscosity of Newtonian fluids. These two laws alone do not result in conflicting modeling requirements, but combined with other laws they often do. Dropping one of the two laws is frequently the only way to resolve the conflict. For instance, turbulent motions of fluid flow are governed much more strongly by inertial forces than by viscous forces; hence, the law of viscous friction may be dropped. Conversely, in the regions of laminar motion of a fluid, inertia has such little effect that Newton's law of inertia can be disregarded.

Deciding whether the flow is laminar or turbulent can usually be based on the magnitude of the Reynolds number. Consider the steady-state flow in a pipe. The flow is governed by viscous and inertial forces and, hence, by the two principal pi-numbers, $\underline{Ne} = F/(\rho \ell^2 v^2)$ and $\pi_v = F/(\mu \ell v)$, as explained in the section "Principal pi-numbers." From these two numbers the Reynolds number evolves as $\underline{Re} = \pi_v /\underline{Ne} = \ell v/\nu$, where $\nu \equiv \mu/\rho$ is the kinematic viscosity. The parameter of interest in pipe flow is the pressure loss, p. The representative pressure force, $F \triangleq p\ell^2$, can be combined with the representative inertial force, $F \triangleq \rho \ell^2 v^2$, to form the principal pi-number $\underline{Eu} = p/\rho v^2$, called the Euler number. In pipe flow, the Euler number is traditionally expressed as the friction coefficient,[1] f, defined as $f = \Delta p/(\frac{1}{2}\rho \bar{v}^2)(d/L)$, where Δp is the pressure drop in the pipe length, L; d is the pipe diameter; ρ is the fluid density; $\bar{v}$ is the mean flow velocity.

Nikuradse[2] published the results of model tests of pipe flow with surface roughness, which he produced with sand grains of uniform diameter, e. He showed that, in general, the friction coefficient, f, is a function of the Reynolds number and the relative roughness, e/d (a common pi-number). At high Reynolds numbers, however, the friction coefficient becomes independent of the Reynolds number, as shown in Fig. 21. Consequently, at high Reynolds numbers, f becomes constant; that is, the representative pressure force becomes proportional to the representative inertial force. This means that the influence of viscosity is negligible.

---

[1] Also known as Darcy friction factor.

[2] J. Nikuradse, "Stroemungsgesetze in rauhen Rohren," (The laws of flow in rough pipes), *Forschungsarbeiten im Ingenieurwesen*, No. 361 (1933).

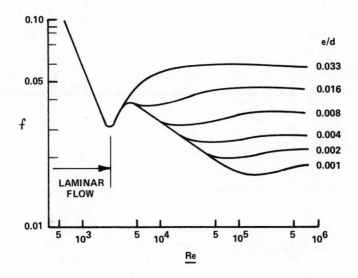

Fig. 21. Nikuradse's "harp" -- friction coefficient of pipe flow.

Figure 21 shows also that for very low Reynolds number, f becomes inversely
proportional to $\underline{Re}$. Since f is the ratio of representative pressure force
to representative inertial force, and $\underline{Re}$ the ratio of representative inertial
force to representative viscous force, the relation of $f \propto 1/\underline{Re}$, valid in the
low $\underline{Re}$ range, means that the representative pressure force becomes proportional
to the representative viscous force. Hence, the influence of inertia is negligible.

An application of relaxation based on a flow diagram shown in Fig. 21, is
given in the case study "Urban Air Pollution." Another well-known example
of conflict arising from scaling both inertial and viscous forces is the
testing of aircraft in a wind tunnel. In general, aircraft performance
is governed by Newton's law of inertia, the law of viscous friction, and the
law of adiabatic air compression. The first two laws lead again to the Rey-
nolds number, $\underline{Re} = \ell v / \nu$. The last law is usually expressed as $p = \rho a^2 / \gamma$
(where a is velocity of sound in air, and $\gamma = c_p / c_v$), a relation that can be
converted into the following principal pi-number

Adiabatic pressure force

$$F \triangleq p \ell^2 \longrightarrow F \triangleq \frac{\rho a^2}{\gamma} \ell^2 \longrightarrow \pi_p = \frac{F \gamma}{\rho a^2 \ell^2}$$

$$p \triangleq \rho a^2/\gamma$$

The Reynolds number is composed of two principal pi-numbers, $\underline{Ne}$ and π_v , as discussed earlier. Similarly, $\underline{Ne}$ can be combined with π_p to form the Mach number.

Mach number[1] $\underline{Ma} = \sqrt{\dfrac{\pi_p}{Ne}} = \dfrac{v}{a} \sqrt{\gamma}$

The identity of the two principal pi-numbers, $\underline{Re}$ and $\underline{Ma}$, would result into the following model rules

$$\frac{v}{v'} = \frac{\ell'}{\ell} \frac{\nu}{\nu'} \quad \text{and} \quad \frac{v}{v'} = \frac{a}{a'} \sqrt{\frac{\gamma'}{\gamma}}$$

If, as usual, air is used in both model and prototype, these two model rules result into two conflicting speed requirements -- $v/v' = \ell'/\ell$, and $v/v' \triangleq 1$, an impossibility in scaling. Therefore, relaxations are intro-duced. In the subsonic flight regime of less than half the speed of sound (approximately 170 m/s), where the effect of air compression is negligibly small, the law of adiabatic air compression and, hence, the Mach number become insignificant, and only the laws of inertia and viscous friction, i.e., $\underline{Re}$, dominate. Accordingly, the model's velocity must be made (ℓ/ℓ') times greater than the prototype's, with the result that the model's velo-city rises well above the speed of sound for all but the slowest planes, and the Mach number, which was negligible in the subsonic prototype, becomes significant in the model.

Theoretically, this difficulty could be overcome by using a fluid of small kinematic viscosity, such as water. But the fluid would have to flow at very high speed, a formidable technical problem. Another solution would be to decrease the kinematic velocity of air by pressurizing it. The increased

---

[1]Usually, γ is omitted so that the Mach number proper appears as the ratio of gas speed to the speed of sound. But then the additional condition of $\gamma = \gamma'$ must be observed.

pressure would increase the density of the air without much effect on either
the speed of sound, a, or the viscosity, μ. Consequently, the Reynolds and
Mach numbers could be satisfied simultaneously if

$$\nu = \nu' \quad \text{and} \quad \ell/\ell' = \rho'/\rho$$

This solution is often impractical because in large structures like wind
tunnels, high pressure requires heavy and expensive construction. Fortunately, there are other relaxations, discussed later in "Law Simulation."

The foregoing example illustrates a very important point in relaxation:
before disregarding a law, the experimenter must satisfy himself of its
insignificance not only to the prototype but to the model, as well. Too
often, a law of little importance to the prototype assumes great importance
when applied to the model. Consider for instance the modeling of hydraulic
phenomena. In full-scale hydraulic structures, surface tension is certainly
insignificant. But with decreasing size, surface tension decreases more
slowly than any other governing effect. Therefore, in very small models,
like harbors and estuaries, surface tension is no longer insignificant.
Special measures to offset this unwanted effect will be presented in "Segmented Modeling."

As pointed out, the technique of disregarding a weak law is perhaps the most
common of all relaxation methods. In low-temperature phenomena, for instance, where radiation plays only a small part, the law of heat radiation
can often be safely neglected. In some types of vibrating structures,
gravity exerts so little influence on the natural frequency that the law of
gravitation need not be considered. In colloidal material, the inner surfaces,
and with them their surface forces, are so large that the weight of the small,
submicroscopic particles can be neglected, and so on. The case studies presented in Part II offer many more examples.

Self-modeling Tests

In self-modeling, the prototype itself becomes the model; here the length
scale factor is clearly unity and the materials are the same because model

and prototype are the same. Any quantity other than length and material
properties, however, can be changed in accordance with the ruling pi-numbers.

Self-modeling experiments are frequently used in fundamental heat and mass
transfer experiments to clarify the influence of certain laws whose parameters
can be varied without affecting length and materials properties. The case
study "Induction Furnace" is a good example. Here, through self-modeling,
a great number of weak laws could be identified.

Self-modeling tests often permit us to establish the governing laws of a
given machinery. Consider for instance the laws governing the performance
of torque converters for automobiles, as described by Ishihara and Emori.[1]
It was argued that a hydraulic transmission is principally governed by
inertial forces of both the working fluid and the moving mechanical parts;
viscous forces were surmised to be negligible because the flow in the hydraulic circuit was thought to be turbulent. The correctness of this assumption
was defended in a self-modeling test.

The two laws possibly governing the torque converter's performance are
Newton's law of inertia and the law for viscous friction of a Newtonian
fluid. Therefore, two principal pi-numbers apply.

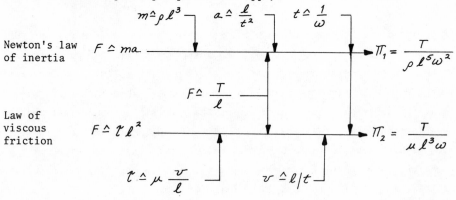

[1]T. Ishihara and R.I. Emori, "Torque converter as a vibration damper and
its transient characteristics," *SAE Trans*., 75, Pt. 2 (1967), 501-512.

where T is torque and ω is circular frequency. For self-modeling tests, $\ell = \ell'$, $\mu = \mu'$, $\rho = \rho'$. Therefore, π_1 deteriorates into the simple model rule of $T/\omega^2 =$ const, and π_2, into the simple model rule of $T/\omega =$ const. If, indeed, viscous forces would play only a minor role, then, by keeping the value of T/ω^2 constant (while changing both T and ω), we would expect all common pi-numbers such as ω_2/ω_1 or T_2/T_1 (subscripts 1 and 2 for pump and turbine) to remain constant, too. If, on the other hand, viscous forces would play a major role, we would expect similarity by keeping T/ω constant. And if both viscous and inertial forces would participate to the same degree, no similarity could be expected.

Tests were run at a special facility, with the pump connected to a driving motor, and the turbine to a dynamometer, as indicated in Fig. 22. The pump speed, ω_1, was kept constant, and pump and turbine torques, T_1 and T_2, were measured at incrementally changed levels of turbine speed, ω_2. Experiments were repeated at three constant pump speeds, $\omega_1 = 800$ rpm, 1100 rpm, and 1400 rpm. Results are plotted in Fig. 22, with ω_2/ω_1 and T_2/T_1 as common pi-numbers and T_1/ω_1^2 representing the principal pi-number derived from Newton's law. The data taken at various ω_1-values all collapse into two curves, as anticipated. Hence, inertial forces do indeed govern the performance of the tested hydrodynamic transmission, and viscous forces can be ruled out.

Segmented Modeling

In many modeling cases, we cannot disregard a governing law throughout the entire range of investigation. But we can often divide the given phenomenon into independent segments, each governed by fewer laws than the whole. Segmented modeling is feasible if a phenomenon changes its behavior as time passes; or if it is composed of spatially separated regions; or if it displays different behavior in different directions or at different speeds. Total performance can be restored by adding or superimposing partial performances.

Sequential modeling. Sequential modeling plays a major role when one phase
follows another. In modeling automobile crashes, for instance, the crash
phase can be clearly distinguished from the post-crash phase. During the
crash phase, which usually lasts only for a fraction of a second, tire
friction forces are negligibly small compared to the large inertial forces
of the impacting cars. Hence, the crash phase is governed by inertial
forces and by energy dissipation in partially elastic and imperfectly smooth
bodies. Tire friction forces do not come into play until the beginning of
the post-crash phase, which may last for several seconds. But then energy
dissipation due to impact need no longer be modeled. Thus, dividing the
whole phenomenon into two sequential phases greatly relaxes the modeling
requirements.

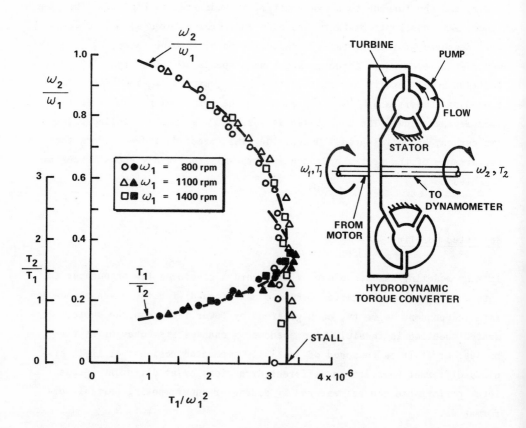

Fig. 22.   Steady-state performance of a hydrodynamic torque converter -- an example of a self-
modeling test.

Non-steady heat transfer is another example. The phenomenon is difficult to model because input energy must be scaled along with stored, conducted, and radiated heat energies. Modeling is sometimes possible, however, if the period of energy input is short and the initial temperature is low. Total performance can then be estimated by separately modeling (a) the initial period of energy input, when almost all heat is stored and conducted but very little heat is rejected, and (b) the final steady-state stage, when the amount of stored heat no longer changes and need not be modeled. In stage (a), only the laws of heat absorption and heat conduction need be considered; the laws of heat radiation and convection can be neglected. In steady state, stage (b), they cannot be neglected; but then, the temperature remains constant, and the law of heat absorption can be disregarded. An example of stage (b) is discussed in the case study "Equilibrium Temperature of a Large Tire."

Regional Modeling. If a phenomenon can be treated as an assembly of separate regions, each governed by its own set of laws, it can be modeled regionally.

An example of regional modeling in the field of mechanical engineering is the dynamic response of a wheeled vehicle traversing undulating terrain. In a slowly moving off-road vehicle, vibrations are confined to frequencies between 1 and 10 Hz. At these low frequencies, sprung and unsprung masses behave like rigid bodies; consequently, those masses are subjected only to Newton's laws of inertia and gravitation. Furthermore, all springs having negligibly small mass follow only Hooke's law of elasticity, and all shock absorbers also having negligibly small mass follow only the law of viscous friction. Hence, the whole vehicle can be partitioned into separate regions where different laws apply. Under these circumstances, modeling is considerably relaxed.

In one well-known example of hydraulics -- the turbulent flow of a fluid through a pipe or channel -- inertial effects are confined to the main stream whereas viscous effects are confined to the vicinity of the walls. This difference permits a number of relaxations. Because the major effects of viscous friction are restricted to the vicinity of the walls (where a large velocity gradient exists), the law of viscous friction need not be expressed in truly general representative terms that would apply to any conceivable

situation. Instead, it can be specified in terms of the boundary layer in
which viscous friction takes place. The viscous shear stress is then
$t_v \triangleq \mu v/\delta$, where δ is the boundary layer thickness. The shear area can
be expressed as $A_w = w l$, where A_w is the wetted wall area, w is the wetted
wall perimeter, and l is the length representing all lengths except the wetted
wall perimeter. Consequently, the pi-number for viscous shear in the vicinity
of the wall, π_w, assumes the following form

$$F \triangleq t_v \, A_w \longrightarrow F \triangleq \mu w l \frac{v}{\delta} \longrightarrow \pi_w = \frac{F \delta}{\mu w l v}$$

$$t_v \triangleq \mu \frac{v}{\delta} \qquad A_w \triangleq w l$$

The law of inertia is applied only to the main stream.

$$F \triangleq m a \longrightarrow F \triangleq \rho A v^2 \longrightarrow \pi_i = \frac{F}{\rho A v^2}$$

$$m \triangleq \rho A l \qquad a \triangleq \frac{v}{t} \qquad t \triangleq \frac{l}{v}$$

A is the cross sectional area of flow.

From π_w and π_i , a pi-number similar to the Reynolds number can be formulated

$$\pi = \frac{\pi_w}{\pi_i} = \frac{\delta v}{l v} \frac{A}{w}$$

We now apply this pi-number to two pipes of the same length carrying the same
fluid at the same mean speed. Consequently, $l = l'$, $v = v'$, $\delta = \delta'$, $\nu = \nu'$.
Similarity of the two flows is then secured if $A/w = A'/w'$. The ratio
is known as the hydraulic radius, R_h.

The concept of hydraulic radius is widely used in hydraulics, particularly
for ducts and open channels with negligible secondary flow. It permits
relaxation of geometric similarity for the cross section of the flow, as
long as the flow is turbulent and the boundary layer is laminar. Under
these circumstances, the cross-sectional shape of the model may deviate from
that of the prototype as long as the hydraulic radius is scaled in accordance with the length scale factor.

Another case of regional modeling is found in the testing of ship models.
The resistance of a ship as it moves through water is made up of two major
components: friction on the hull, and wave-making resistance. These two
components are governed by inertia, viscosity, and gravity, and hence, by

Newton's law of inertia, by the law of viscous friction of a Newtonian
fluid, and by the law of gravitation. The first two can be expressed in
terms of the Newton number, $\underline{Ne} = F/(\rho\,\ell^2 v^2)$, and the principal pi-number
of $\underline{\pi_v} = F/(\mu\,\ell\,v)$, as derived in the section "Principal and Common Pi-numbers."
The gravitational force yields the pi-number of $\pi_g = F/(\rho\,g\,\ell^3)$, as shown in
the section "Why Relaxation?" The first two pi-numbers are combined to the
Reynolds number, $\underline{Re} = v\ell/\nu$, the first and the last to the Froude number,
$\underline{Fr}^2 = \pi_g/\underline{Ne} = v^2/(g\,\ell)$. To achieve similarity, model and prototype must be
operated at equal Froude and Reynolds numbers, an impossibility if both
model and prototype are tested in the same fluid because $\underline{Re}$ demands that
$v/v' = \ell'/\ell$, and $\underline{Fr}$ that $v/v' = (\ell/\ell')^{1/2}$. If a different fluid is used for the
model, its kinematic viscosity must be reduced by 3/2 power of the length
scale factor.

$$\underline{Re} \longrightarrow \frac{v}{v'} = \frac{\nu}{\nu'}\frac{\ell'}{\ell}$$

$$\underline{Fr} \longrightarrow \frac{v}{v'} = \sqrt{\frac{\ell}{\ell'}} \qquad \Bigg\} \longrightarrow \frac{\nu}{\nu'} = \left(\frac{\ell}{\ell'}\right)^{3/2}$$

Even with a length scale factor as small as 100, a ratio that produces
undesirably large models, the required kinematic viscosity of the model
fluid would be 1/1000 that of water. Such a fluid does not seem to exist.

Because of these conflicting scaling requirements, hull friction and wave
resistance cannot be modeled simultaneously. However, with friction
effects confined predominantly to the hull's boundary layer, and gravity and
inertia effects restricted to waves around the ship, it is argued that the
two resistances can be determined separately in the following procedure
(originally suggested by Froude).

The total resistance, R_T, is divided into two parts, the hull friction, R_F,
and the wave-making resistance, R_R, so that $R_T = R_F + R_R$. The hull friction,
R_F, is taken to be a function of only the Reynolds number (i.e., of viscous
and inertial forces); and the wave-making resistance, R_R, of only the Froude
number (i.e., of inertial and gravitational forces). The model is then run
at a speed that complies with the Froude number, i.e., $v/v' = \sqrt{\ell/\ell'}$, and the
total model resistance, R_T', is measured. The frictional resistance of the
model at this speed is calculated and subtracted from the total resistance

to yield the wave-making model resistance $R'_R = R'_T - R'_F$. The calculation
of the frictional resistance is based on the assumption that it is that of
a flat plank of the same length and surface area as the model. With the
wave-making resistance of the model isolated, the wave-making resistance of
the prototype is calculated from the Newton number: $R_R = R'_R \ell^2 v^2 / (\ell'^2 / v'^2)$,
with $v / v' = \sqrt{\ell / \ell'}$. The frictional resistance of the prototype, R_F, is again
calculated by using the plank approach, and added to R_R. The result is
the total resistance of the prototype. With modifications and refinements,
this method is still used in all model basins.

As a last example, consider the steady-state heat flow through a wall whose
inner side is heated and kept at a constant temperature and whose outer
side radiates heat at a constant but lower temperature (such as a spacecraft). Heat flow through the wall is governed by the law of heat conduction;
heat rejection at the outer wall by the law of heat radiation. From these
two laws, two principal pi-numbers evolve.

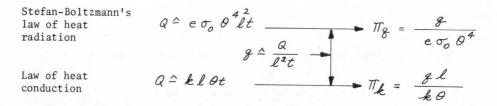

Stefan-Boltzmann's law of heat radiation
$$Q \triangleq e\sigma_o \theta^4 \ell t$$

$$q \triangleq \frac{Q}{\ell^2 t}$$

$$\pi_q = \frac{q}{e\sigma_o \theta^4}$$

Law of heat conduction
$$Q \triangleq k\ell\theta t$$

$$\pi_k = \frac{q\ell}{k\theta}$$

where e is the emissivity of gray body, σ_o is the Stefan-Boltzmann constant,
and k is the thermal conductivity.

Suppose model and prototype were to be manufactured from the same material
and operated at the same surface temperature. The input heat flux would then
scale according to π_q as $q / q' = 1$, and according to π_k as $q / q' = \ell / \ell'$ -- clearly
two conflicting results. Regional modeling, however, offers relief. Since
each of the two laws is acting within its own region -- the law of heat
radiation, outside the wall; and the law of heat conduction, within the
wall -- two temperature scale factors can be employed -- one for the temperature drop across the wall thickness, $\Delta\theta$, and one for the surface temperature,
θ_s. Then, with $\theta_s = \theta'_s$,

$$\Pi_g \longrightarrow \frac{g}{\theta_s^4} = \frac{g'}{\theta_s'^4} \longrightarrow \frac{g}{g'} = 1$$

$$\theta_s = \theta_s'$$

$$\Pi_k \longrightarrow \frac{g\ell}{\Delta\theta} = \frac{g'\ell'}{\Delta\theta'} \longrightarrow \frac{\Delta\theta}{\Delta\theta'} = \frac{\ell}{\ell'}$$

Hence, regional modeling offers considerable experimental advantages; it permits manufacturing of both model and prototype from the same material and running them at the same surface temperature, provided the same input heat flux is applied. As a result, the temperature drop across the wall can be expected to scale in agreement with the length scale factor.

Directional Modeling. When governing laws work in distinct directions instead of distinct regions, directional modeling[1] can relieve stringent scaling requirements. For example, if the turbulent flow in rivers and estuaries were faithfully scaled in horizontal dimensions as well as in depth, river models would be so shallow that turbulent flow would be suppressed and unwanted surface tension would be exaggerated. With directional modeling, the depths of river and estuary models are very often disproportionally magnified or "intentionally distorted" geometrically for the following reasons: fully turbulent flow in a sloped channel is governed by Newton's law of inertia and the law of gravitation; but the inertial forces act predominantly in a horizontal direction, while the gravitational forces

---

[1]Directional modeling has been treated thoroughly by H.E. Huntley in *Dimensional Analysis*, Rinehart, New York, 1951. Huntley adapts directional modeling to dimensional analysis by resolving a vector length into three mutually orthogonal components, an increase in the number of fundamental quantities from one to three. Judiciously applied, the three directional lengths provide more power to dimensional analysis than the single "scalar" length, as Huntley shows in many examples.

See also the interesting discussion by F.V. Costa, "Directional analysis in model design," *Proc. ASCE, J. Engineering Mech. Div.* 97, EM 2, 519-539 (Apr 1971). Also, J. Gessler, "Vectors in dimensional analysis," *Proc. ASCE, J. Engineering Mech. Div.* 99, EM 1, 121-129 (Feb 1973).

We may note that the physical reasons for directional modeling are more clearly manifest in the law approach pursued here than in the formal approach of dimensional analysis.

act in a vertical direction. Considering a cross-sectional segment of channel flow (Fig. 23), we find its weight represented by $\rho g z \ell^2$, where the length ℓ stands for horizontal dimensions, z for vertical dimensions, g for gravitational acceleration, and ρ for fluid density. Likewise, its mass is represented by $\rho z \ell^2$, and its speed (in flow direction), by ℓ/t.

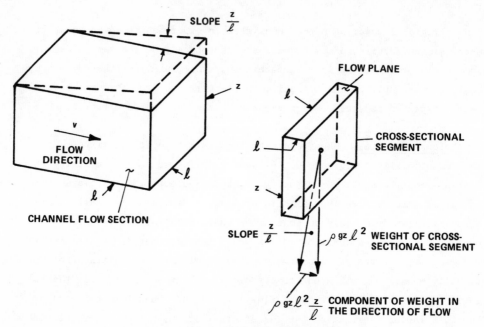

Fig. 23.   Weight components in channel flow
z = representative vertical length
ℓ = representative horizontal length

The flow is sustained by the weight component in the direction of flow. The component derives from the law of gravitation and is expressed as

$$F_{flow} \triangleq \rho g z \ell^2 \frac{z}{\ell} \longrightarrow \pi_g = \frac{F_{flow}}{\rho g z^2 \ell}$$

Inertial forces in the direction of flow are derived from Newton's law of inertia and are expressed as

$$F_{flow} \triangleq \rho z \ell^2 \frac{\ell}{t^2} \longrightarrow \pi_i = \frac{F_{flow} t^2}{\rho z \ell^3}$$

Both pi-numbers are valid only in flow-direction. Their ratio yields
another pi-number

$$\pi = \frac{\pi_i}{\pi_g} = \frac{g\,z\,t^2}{\ell^2} \quad\xrightarrow{\;v \triangleq \frac{\ell}{t}\;}\quad \pi = \frac{g\,z}{v^2}$$

Hence, flow conditions of model and prototype become similar if, under the
same gravitational acceleration, $z/z' = (v/v')^2$. The velocity, v, can be
replaced by the average velocity computed from the rate of discharge,
$\dot{Q} \triangleq v\ell z$. Then,

$$\pi \xrightarrow{\;v \triangleq \dot{Q}/\ell z\;} \frac{g\,z^3\ell^2}{\dot{Q}^2} \xrightarrow{\;g = g'\;} \frac{\dot{Q}}{\dot{Q}'} = \frac{\ell}{\ell'}\left(\frac{z}{z'}\right)^{3/2}$$

Note that all these relations are true only for the direction of flow.

The developed pi-number for channel flow allows for a large difference
between the horizontal and vertical length scale factor. Models with a depth
exaggeration of 40 or more have been tested. Since the slope is represented
by z/ℓ, the slope in the model is also exaggerated. An illustration is
given in the case study "Sedimentation: Sediment Waves."

Directional modeling can always be applied when a law acts predominantly
along a given direction -- heat conduction, capillary action, elastic
deformation, and channel flow. For directional elastic deformations, a
special principal pi-number has been introduced -- the Strouhal number.
Suppose an elastic system like a bridge or an aircraft wing must be modeled
with respect to its elastic vibrations in a given direction (vertical, horizontal, rotational). The elastic forces can then be expressed by a spring
constant, k, valid for the given direction, so that Hooke's law assumes the
simple representative form of $F \triangleq k\ell$. The inertial forces of the system
follow Newton's law of inertia expressed representatively as $F \triangleq m v^2/\ell$. With
these two relations, the principal pi-number of $\pi = \frac{\ell}{v}\sqrt{\frac{k}{m}}$ can be formed.
Since for an elastic system the ratio of $\sqrt{\frac{k}{m}}$ represents the natural frequency ω, this pi-number can be expressed as $St = \ell\omega/v$, known as the Strouhal
number.

Another principal pi-number based on directional modeling is the Dean number, which is a special version of the Reynolds number, defined for flow in

curved pipes. On a curved path, the fluid is exposed to two kinds of mass forces: the inertial forces of straight flow, and the centrifugal forces resulting from the pipe curvature. In a strict representative sense, both forces are covered by just one of the many expressions developed earlier for Newton's law of inertia -- for instance, by $F \triangleq \rho \ell^2 v^2$. Here, however, practical reasons called for restricting the representative character of this expression. The Dean number was devised to indicate the influence of the radius of curvature on secondary flow patterns. Hence, two representative lengths were specified, representing the directions of the two inertial forces -- radius of curvature, R, indicating the direction of centrifugal forces, and radius of the pipe's cross section, r, representing all other lengths besides R and indicating the rest of the inertial forces. With these specifications, two formulations of Newton's law of inertia are possible

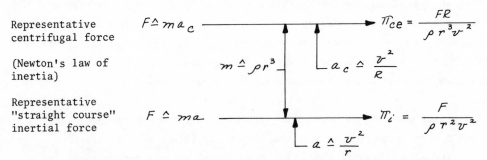

| Representative centrifugal force | $F \triangleq m\, a_c$ | $\pi_{ce} = \dfrac{FR}{\rho\, r^3 v^2}$ |

(Newton's law of inertia) $m \triangleq \rho r^3$ $a_c \triangleq \dfrac{v^2}{R}$

| Representative "straight course" inertial force | $F \triangleq m a$ | $\pi_i = \dfrac{F}{\rho\, r^2 v^2}$ |

$a \triangleq \dfrac{v^2}{r}$

In addition to these two inertial forces, the viscous forces present are representatively expressed by

Law of viscous friction of Newtonian fluid $F \triangleq \ell\, r^2$ $\pi_v = \dfrac{F}{\mu\, r v}$

$\ell \triangleq \mu \dfrac{v}{r}$

The Dean number is defined as

$$\underline{De} \;=\; \frac{\pi_v}{\sqrt{\pi_{ce}\,\pi_i}} \;=\; \frac{r v}{\nu}\sqrt{\frac{r}{R}}$$

For flow between two rotating concentric cylinders, the pipe radius r (representing all lengths except the radius of curvature) can be replaced by the
width of the annulus, b. The velocity, v, can be replaced by the product of
angular velocity ω and the mean radius of the annulus, R_a; and the radius
of curvature, by R_a. For this special application, the Dean number is
occasionally called the Taylor number

$$Ta = \frac{\omega \sqrt{R_a\, b^3}}{\nu}$$

The author of this book[1] performed a model experiment in which a combination
of regional and directional modeling was used to resolve conflicting claims
of Re and Fr in a seepage problem. In general, flow through porous media
is governed by viscous friction between fluid and grains, inertial forces
of the fluid, and fluid weight. Hence, correct modeling requires observation
of the Reynolds and the Froude numbers, as demonstrated in the section on
"Regional Modeling." If the same fluid is used in the model as in the prototype (i.e., if $\nu = \nu'$), then modeling without drastic relaxations becomes
impossible. But if different fluids are used, then it would be necessary
that $\nu/\nu' = (\ell/\ell')^{3/2}$, a requirement practically impossible to meet for large
scale factors. Here, the problem was solved by employing two different
length scale factors.

In the tests, sand of uniform grain size was held in the bottom of a glass
tube by a thin cloth with negligible flow resistance; when the tube was
filled with water, gravity made the water flow downward through the sand
(Fig. 24). Tests were performed with sands of three different grain sizes,
with the depth of the sand layer made proportional to the grain size to
maintain geometrical similarity.

Because the effects of gravitation depends on the water column's vertical
height, z (water region), while the inertial and viscous effects are confined to the layer of sand of the representative dimension, ℓ (sand region),
we can describe seepage by two representative lengths, z and ℓ. Note that
ℓ represents the dimensions of the sand layer as well as grain and void
sizes.

---

[1]Together with R.I. Emori.

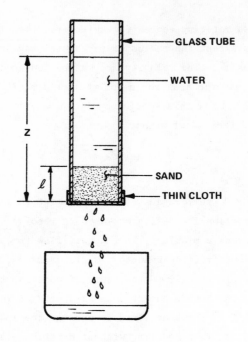

Fig. 24.   Seepage model experiment.

With these relaxations, we can rewrite the representative versions of the three governing laws as follows:

Newton's law of inertia for fluid flow in sand layer

$$F \triangleq ma \qquad\qquad\qquad\qquad \pi_i = \frac{F}{\rho A v^2}$$

$$m \triangleq \rho A l \qquad a \triangleq \frac{v^2}{l}$$

Law of viscous flow of Newtonian fluid

$$F \triangleq \tau A \qquad\qquad\qquad\qquad \pi_v = \frac{F l}{\mu A v}$$

$$\tau \triangleq \mu \frac{v}{l}$$

Law of gravitation for water column

$$F \triangleq mg \qquad\qquad\qquad\qquad \pi_g = \frac{F}{\rho g A z}$$

$$m \triangleq \rho A z$$

A is the representative area.

From the three resulting principal pi-numbers, two others can be abstracted
by eliminating F. From them, two model rules evolve

$$\pi_1 = \frac{\pi_v}{\pi_i} = \frac{\ell \rho v}{\mu} \longrightarrow \ell v = \ell' v'$$

(Reynolds no.)

$$\rho = \rho' \quad \Big] \quad \text{same}$$
$$\mu = \mu' \quad \Big] \quad \text{fluid}$$

$$\pi_2 = \frac{\pi_i}{\pi_g} = \frac{g z}{v^2} \longrightarrow \frac{z}{v^2} = \frac{z'}{v'^2}$$

(Froude no.)

With the same fluid in all tests, plots of z/v^2 vs ℓv are essentially those
of $\underline{Fr}$ versus $\underline{Re}$. If, in this plot, $\underline{Fr}$ should turn out to be inversely proportional to $\underline{Re}$, as shown in Fig. 25; i.e., if $\frac{\pi_v}{\pi_i} \propto \frac{\pi_g}{\pi_i}$, then inertial
effects are shown to be negligible. If, on the other hand, $\underline{Fr}$ becomes con-
stant as $\underline{Re}$ varies, then viscous effects are shown to be negligible. If
$\underline{Fr}$ should cover the "knee" in Fig. 25, then neither viscous nor inertial
effects are negligible; instead, both are important.   (See Problem 14,
Appendix A.)

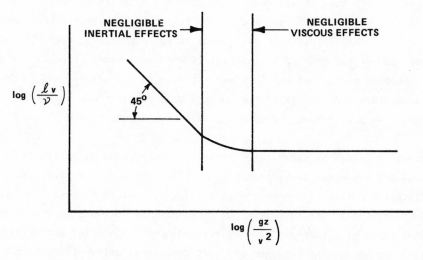

Fig. 25.   Inertial and viscous effects at low and high fluid velocities.

Circumventing Strong Laws

If none of the conflicting laws can be disregarded, we must abandon the idea
of modeling the whole phenomenon faithfully. Instead, we can try to circumvent the most troubling law by any one of the following devices.

Restrictions in Generality

The phenomenon is broken up in a number of special cases, each governed by
fewer laws than the entire phenomenon. If enough special cases are investigated, and if they are all relatively independent of each other, an approximation of the total phenomenon can be obtained by superimposing results of
the special cases.

As an example, consider a ship in a severe storm at sea, a phenomenon sufficiently complex to warrant modeling special cases. Since the ship is operating in two distinctly different environments (water and wind), we may test
two special cases: the ship in still water but exposed to wind forces, and
the ship in still air but exposed to water forces. Another special case that
can be investigated independently of water and wind resistance is the hull's
structural response to wave impact. Each of these special cases allows more
scaling freedom than is allowed by the phenomenon in its entirety. As a
corollary, those effects that are not pertinent to the special case need not
be modeled. That is, if wave resistance is to be measured, the structural
vibrations may not have to be modeled; if wind resistance is to be measured,
the submerged portion of the ship may not have to be modeled; and so on. But
to superimpose and synthesize the special-case results into a picture of
the total ship performance requires great care and experience.

Modeling soil-machine interaction presents another case in point. Most
natural soils possess characteristics of both clay and sand; i.e., cohesion
and internal friction. These characteristics cannot be modeled simultaneously,
as demonstrated in the case study on "Soil-Machine Interactions;" but the
special soils of either dry sand or soft clay may be modeled separately
because dry sand lacks cohesion and soft clay lacks internal friction. Hence,
by testing the model of an earth-working device first in dry sand and then in

soft clay, we can estimate with some confidence its performance in mixed soils
having both cohesion and friction.

Law Simulation

Sometimes a disturbing but indispensible law can be substituted by a device
that would simulate the desired effects and bypass the undesirable ones. Of
the many possibilities, three are most frequently employed: dummy weight,
transition fixing, and dummy springs.

Dummy Weights. Many large elastic structures are deformed not only by
external forces but also by their own weight. Modeling requires consideration of at least two laws: the law of gravitation and Hooke's law of elastic
deformation. The two ensuing principal pi-numbers are:

Law of
gravitation

$$F \triangleq mg \longrightarrow F \triangleq \rho g \ell^3 \longrightarrow \pi_g = \frac{F}{\rho g \ell^3}$$

$$m \triangleq \rho \ell^3$$

Hooke's law
(disregarding
lateral con-
tractions)

$$F \triangleq \sigma A \longrightarrow F \triangleq E \epsilon \ell^2 \longrightarrow \pi_e = \frac{F}{E \epsilon \ell^2}$$

$$\sigma \triangleq E \epsilon \qquad A \triangleq \ell^2$$

From π_g and π_e, a third pi-number can be formulated

$$\pi = \frac{\pi_g}{\pi_e} = \frac{E \epsilon}{\rho g \ell} \longrightarrow \frac{\ell}{\ell'} = \frac{E/\rho}{E'/\rho'}$$

Geometrical $\epsilon = \epsilon'$ $g = g'$
similarity

This pi-number constitutes an obstacle to large length scale factors. The
ratio of E/ρ represents the square of the velocity of sound (in rods), a_s,
so that $\ell/\ell' = (a_s/a_s')^2$. For metals the velocity of sound does not change much.
If the prototype is of steel, for instance, even the choice of an exotic
material like silver allows for a length scale factor of only 3.5 -- too
small for practical purposes.

The limitations imposed by the velocity of sound can be relaxed, however, by increasing the weight of the model structure without changing its elastic properties. This is done with a series of equally and closely spaced dummy weights attached rigidly to the model structure. Now each section of the structure, with its extra weight, can be considered to possess a density of

$$\rho' = \rho_0' + \frac{\Delta m'}{\Delta V'}$$

where ρ_0' is the real density of the model structure, $\Delta V'$ is the volume per dummy weight, and $\Delta m'$ the dummy weight's mass. Thus, the original restricting length scale factor relation becomes more flexible,

$$\frac{\ell}{\ell'} = \frac{E}{\rho} \; \frac{\rho_0' + \Delta m' / \Delta V'}{E'}$$

Transition Fixing. As pointed out in the section on "Disregarding Weak Laws," aerodynamic model tests are usually run at speeds lower than required by the Reynolds number because, otherwise, the model's velocity rises well above the speed of sound for all but the slowest planes, and effects arising from the compressibility of air become disturbingly important.

The ill effects of running the model at a lower Reynolds number than the prototype are compensated by "transition fixing." Experiments and theory have shown that the difference of flow patterns and, consequently, of drag and lift forces at different Reynolds numbers originate in the boundary layer adjacent to the wall surface. In particular, at higher Reynolds numbers the boundary layer changes from the laminar to the turbulent state thereby for blunt bodies sharply reducing the drag. As originally observed by Prandtl,[1] a laminar boundary layer can be made turbulent by small surface disturbances. Accordingly, bands with rough surfaces are added to certain areas of the model aircraft to initiate transition to the turbulent flow that would occur at the higher Reynolds number of the prototype.[2]

---

[1] L. Prandtl, "Über den Luftwiderstand von Kugeln" (On the air resistance of spheres), *Göttinger Nachr. Math.-Phys. Klasse*, 177 (1914).

[2] A.L. Braslow, R.M. Hicks, and R.V. Harris, "Use of grit-type boundary layer transition trips on wind tunnel models," NASA TN-D-3579, Washington, D.C. (Sep 1966). Also, J.C. Daugherty and R.M. Hicks, "Measurements of local skin friction downstream of grit-type boundary layer transition trips..." *AIAA J.* 8, 5, 940-941 (May 1970).

<u>Dummy Springs</u>. For modeling wind-induced vibrations of large elastic struc-
tures such as aircraft or suspension bridges, testing components rather
than the entire model is often justifiable. The component must be scaled
in shape and mass but it can be rigid and it must be supported by springs
that would simulate the elastic interaction of the component with the rest
of the structure. An example is the model investigation performed on a
cable-stayed girder bridge.[1] Several weeks after being opened to traffic
in 1967, the bridge was observed to vibrate vertically in winds between
40 and 50 km/h at a frequency of 0.6 Hz. With the exciting wind speed confined to this narrow band, the major cause was identified as vortex shedding,
and the problem was to find an inexpensive modification of the structure
that would attenuate or eliminate the vibration.

The phenomenon was assumed to be governed by Newton's law of inertia applied
to both bridge and interacting air, Hooke's law applied to the bridge, and
a law describing internal energy dissipation of the bridge. The first two
laws result in the Strouhal number discussed in the section on "Directional
Modeling," $St = \omega \ell / v$, where ω is the frequency, and v is the wind speed.

Consequently, with a length scale factor of 30 and ω = 0.6 Hz, the model
wind speed was $v' = v \omega' / 18$, a relation that permitted choosing the model
frequency ω' at convenience. Therefore, instead of testing a model of the
entire bridge, only a section was exposed to the wind speed v' in a wind
tunnel. An external mounting system provided vertical vibrations of the
sectional model with a selected frequency ω', while restraining torsional
and flexural motions, Fig. 26. With this sectional model, many corrective
modifications of the external aerodynamic shape could be explored.

---

[1] R.L. Wardlaw and C.A. Ponder, "An example of the use of wind tunnels for
investigating the aerodynamic stability of bridges," *Quarterly Bull. Div.
of Mech. Eng. and Natl. Aeronautical Establishment*, Natl. Res. Council
Canada, Rep. No. DME/NAE, 1969 (3) (Jul 1 to Sep 30, 1969), Ottawa, Canada

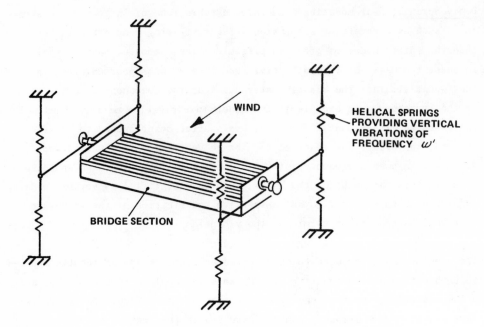

Fig. 26.   Sectional model of a bridge.

Restriction to Gross Effects

Some physical phenomena are formed by the cumulative actions of many small elements whose individual behavior is of no interest because it is the gross effect that we seek. For instance, if we tried to describe the phenomenon of tire friction resulting from the intricate interactions between molecules of rubber, asphalt, water, and air, we would soon bog down. Or, if before constructing a building, we had to predict the bearing capacity of soil in terms of forces and energies between all minute soil particles, we would never be ready to pour the concrete. Although we continue to strive for fundamental insight, we must often go ahead, making use of whatever is at hand.

What is at hand, usually, are the relations among macroscopical quantities. Using these, we can make reliable predictions within practical limits. Moreover, the macroscopical viewpoint often helps us to relax impractical microscopical scaling requirements by modeling the integrated effects of numerous

elements instead of their individual actions. How this works -- in space
as well as in time -- is indicated by the following examples.

Spatially Integrated Effects. In working the soil with bulldozers and tillers,
soil is moved "en masse;" movements of single grains are of little interest.
Thus, model tests need not be concerned with the properties, sizes, and
interactions of single grains. Soil is considered to be properly scaled if
its overall response is physically similar to the prototype, as demonstrated
in the case study on "Soil-machine Interactions."

Sometimes, inconveniently strict model rules can be relaxed by spatial
integration that reduces a three-dimensional phenomenon to a one- or two-
dimensional problem. In modeling the thermal performance of a spacecraft,
heat flow through the outer walls was to be scaled.[1] Because the spacecraft
walls were very thin, local temperatures across the wall could be assumed to
be constant and heat flow could be considered to occur only along the wall.
With this assumption of two-dimensional heat flow, the choice of wall material
and wall thickness could be relaxed.

The concept of spatially integrated effects is particularly useful when
scaling such surface details as chamfer, welding seams, and rivets is unnec-
cessary, as in structural problems. In other instances, wall roughness is
of first importance -- pressure loss in pipes and flow separation on wings
are examples of phenomena that depend primarily on surface roughness. In
these, spatial modeling can be applied; instead of scaling each asperity,
the overall effects of roughness can be generated artificially by means such
as tripping wires on wings and sand particles on pipe walls.

In tests of vehicle tires, true geometrical similarlity would require that
the cords in the plies of the model tire be scaled down. But full-scale
tire cords are already thin, and to make and arrange geometrically thinner
model cords would be tedious indeed. Instead, as described in the case study
on "Equilibrium Temperature of a Large Tire," full-scale cords were used in

---

[1]V.G. Klockzien and R.L. Shannon, "Thermal scale modeling of spacecraft,"
Paper 69 0196 presented at SAE Intl. Automotive Eng. Congr., Detroit,
Mich. (Jan 1969).

the model, for tire performance was judged to be affected by the combined action of many cords.

An interesting combination of spatially integrated and directional modeling occurs in modeling laminar flow through porous material (seepage, sound absorption). When modeling losses of viscous flow in porous material, one need not attempt to scale the pores and capillaries in shape, size, distribution, and direction. Instead, one can assign an effective capillary radius to the material, known as permeability k. To arrive at k, one must take a statistical view and regard pores as parallel capillaries with a representative radius, R, and a representative length, L, Fig. 27. Hence, the porous material is replaced by a battery of parallel capillaries, each with the same radius, R, and each occupying the same area, A, of the total cross sectional area. Since the flow occurs in the direction of L, directional modeling can be applied to each pore alike. The law of viscous friction then takes the form of $F \triangleq tRL$, where RL represents the wall surface of the capillary. The shear stress t can be expressed representatively as $t \triangleq \mu v/R$, where v/R is the representative velocity gradient. The force F is, in terms of the applied pressure, $F \triangleq p R^2$, where R^2 represents the cross-sectional area of the capillary. Finally, the velocity can be replaced by the volumetric rate of flow applied to the area A, $v \triangleq (Q/t)/A$. From all this, the permeability is derived as follows

Law of viscous friction applied to capillary

$$F \triangleq t RL \longrightarrow \frac{Q}{t} \triangleq R^2 \frac{A}{\mu} \frac{p}{L}$$

$$t \triangleq \mu \frac{v}{R} \qquad F \triangleq p R^2 \qquad v \triangleq \frac{Q/t}{A}$$

The gross quantities of flow rate (represented by Q/t), cross sectional area (represented by A), and pressure gradient (represented by p/L) can be measured by subjecting a sample of the given porous material to a flow test. The permeability is then defined as

$$k = \frac{\dot{Q}\mu}{A \, dp/dL}$$

By comparing this definition with the derived representative version of the law of viscous friction, one can easily see that the permeability stands for the representative pore radius, $k \triangleq R^2$. The permeability is expressed in darcy; 1 darcy = 1.02 x 10^{-8} cm^2. An application of the modeling of porous material is given in the case study on "Architectural Acoustics."

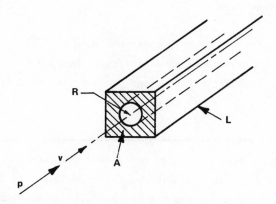

Fig. 27.   Representative capillary pore.

<u>Temporally Integrated Effects</u>. If instantaneous data at a specific time are
not needed -- that is, if only the aggregative results of a phenomenon are
to be modeled, we can find other means of relaxation. The scaling of cumulative effects depends very much on the given phenomenon. For instance,
one model study focused on the kinematics of two cars after impact.[1] During
the collision, the cars exchange momentum so that, at the end of the crash
phase when the cars separate and start following their own trajectories,
each car is provided with a new set of velocities. These velocities, our
sole interest, are fixed by the total momentum exchanged during the crash
phase; therefore, we are not interested in the history of momentum exchange
but only in the end result -- the new set of car velocities. Consequently,
only the end effect of the impact phase (the end momentum of each car) need
be scaled.

The same philosophy obtains for modeling the effects of explosions. The time
history of the explosion itself is of no concern if the following events are
slow. Then, only the total amount of explosive energy need be modeled, as
demonstrated in the case study on "Dynamic Response of Structures -- Explosive Forming of Sheet Metal."

---

[1]R.I. Emori and D. Link, "A model study of automobile collisions," Paper
69 0070 presented at SEA Auto. Engr. Congr., Detroit, Mich. (Jan 1969).

Use of Analytical Knowledge

Scale model experiments are usually performed when analytical knowledge is lacking. However, when at least some analytical knowledge is available, it should be applied, depending on the given problem, as the following example will illustrate.

The problem was to predict the ramming mode performance of an icebreaker ship of the U.S. Coast Guard by model tests conducted in an arctic pool filled with fresh water ice.[1] (The model tests could not be validated, however, because the prototype was not yet built.) In the ramming mode (as distinguished from the continuous mode), an icebreaker progresses through ice-covered water by repeatedly impacting the ice; after each impact, the ship backs away from and then accelerates toward the ice, breaking it in three phases: (1) the ship impacts and crushes the ice, (2) the ship slides upon the ice; (3) the weight of the ship upon the ice causes it to bend and finally to break. In this process, three forces (i.e., three laws) dominate: the inertial forces of the ship, the gravitational forces of the ship, and the ultimate stress forces of the ice. All other forces due to the acceleration and submersion of the broken ice, and to friction and wave making of the ship play only minor roles and can be neglected.

From the three dominating laws, three principal pi-numbers are derived.

Newton's law
of inertia
$$F \triangleq ma \qquad \pi_i = \frac{F}{\rho \ell^2 v^2}$$

$$m \triangleq \rho \ell^3 \qquad a \triangleq \frac{v^2}{\ell}$$

Law of gravitation
$$F \triangleq mg \qquad \pi_g = \frac{F}{\rho g \ell^3}$$

Stress-strain
relation of ice
$$F \triangleq \sigma \ell^2 \qquad \pi_\sigma = \frac{F}{\sigma \ell^2}$$

[1] R.M. White and G.P. Vance, "Icebreaker model tests," *Naval Engrs. J.* 79, 4, 601-605 (Aug 1967). Also, G.P. Vance, "Model tests in ice," *Naval Engrs. J.* 80, 2, 259-264 (Apr 1968).

In the last pi-number, the representative stress, σ, stands for the entire stress-strain relationship of ice. Since, however, only the failure point of the entire stress strain curve is of interest, the representative stress can be interpreted as ultimate stress, σ_u. The ultimate stresses of salt-water ice and fresh-water ice are approximately the same, as are their densities. Thus, if fresh-water ice is used, tests with small models are excluded because, then,

$$\frac{\pi_g}{\pi_\sigma} = \frac{\sigma_u}{\rho g l} \qquad \begin{array}{c} g = g' \\ \sigma_u = \sigma_u' \\ \rho = \rho' \end{array} \Big\} \text{same ice} \longrightarrow l = l'$$

An unsuccessful attempt was made to find other fluids that would produce a much smaller yield stress in the solid phase than water but would have the same degree of tractability and the same low viscosity (so that viscous forces could be neglected). In a situation like this, infusing some analytical knowledge often proves useful. Here, it was assumed that because failure of the ice is mainly ascribed to its being bent, ice will follow the theory of flexure. The bending moment was representatively expressed as $M \triangleq l h^2 \sigma_u$. Note that a distinction was made between horizontal ice dimension, l (which also expressed all ship dimensions), and vertical ice dimension, h. The distinction permitted the simulation, by fresh-water ice, of an effective ultimate stress much lower than the physical ultimate stress. The simulation was accomplished by re-formulating the stress-strain relationship of ice with the help of the postulated representative moment,

Effective
stress-strain
relation of ice
$$F \triangleq \frac{M}{l} \qquad \begin{array}{c} \\ M \triangleq l h^2 \sigma_u \end{array} \Big\} \longrightarrow F \triangleq h^2 \sigma_u \longrightarrow \overline{\pi}_\sigma = \frac{F}{h^2 \sigma_u}$$

This new pi-number, combined with the pi-number developed earlier from the law of gravitation, evolves the relation

$$\frac{\pi_g}{\overline{\pi}_\sigma} = \frac{h^2 \sigma_u}{\rho g l^3} \qquad \begin{array}{c} \sigma_u = \sigma_u' \\ \rho = \rho' \end{array} \Big\} \text{same ice} \quad g = g' \longrightarrow \frac{h}{h'} = \sqrt{\left(\frac{l}{l'}\right)^3}$$

For instance, with $\ell/\ell' = 50$, $h/h' = 354$ -- a result that could be easily implemented in the controlled environment of an arctic pool.

This example, like the previous ones, illustrates clearly that most relaxation techniques can be considered attempts to circumvent the difficulties involved in modeling materials properties. The following example is another case in point.

The most violent impact suffered by a ship at sea is probably the heavy blow taken when the bow re-enters the water. After structural failure was experienced by an aircraft carrier in rough seas off Cape Horn, model tests were conducted to study the vibratory response of the carrier to wave impact loads.[1] The ship's behavior was assumed to be governed by inertial forces of the ship and the water (Newton's law of inertia), by the weight of the ship and the water (law of gravitation), and the elastic forces of the ship (Hooke's law of elasticity). Since propulsion was of no immediate interest, resistance due to fluid friction was neglected.

The three governing laws were transformed into the following principal pi-numbers and model rules.

Newton's law of inertia	$F \triangleq ma$		$Ne = \dfrac{F}{\rho \ell^2 v^2}$
		$m \triangleq \rho \ell^3 \quad a \triangleq \dfrac{v^2}{\ell}$	
Law of gravitation	$F \triangleq mg$		$\pi_g = \dfrac{F}{\rho g \ell^3}$
		$\sigma \triangleq E\epsilon$	
Hooke's law of elasticity	$F \triangleq \sigma \ell^2$		$\pi_e = \dfrac{F}{E \epsilon \ell^2}$

[1] J. Andrews and J.W. Church, "A model for the simulation of wave impact loads and resulting transient vibration of a naval vessel," *Colloquium Use of Models and Scaling in Shock and Vibration*, W.E. Baker (ed), ASME Winter Annual Meeting (Nov 1963), pp. 16-28, ASME, New York.

$$g = g'$$

Froude number $\quad Fr = \sqrt{\dfrac{\pi_g}{Ne}} = \dfrac{v}{\sqrt{g\,\ell}} \quad\longrightarrow\quad \left(\dfrac{v}{v'}\right)^2 = \dfrac{\ell}{\ell'}$$

$$\dfrac{\ell}{\ell'} = \dfrac{E}{E'}\dfrac{\rho'}{\rho}$$

$$\epsilon = \epsilon'$$

Cauchy number $\quad Ca = \dfrac{\pi_e}{Ne} = \dfrac{\rho\,v^2}{E\,\epsilon} \quad\longrightarrow\quad \left(\dfrac{v}{v'}\right)^2 = \dfrac{E}{E'}\dfrac{\rho'}{\rho}$$

Hence, the length scale factor depends on the model's and prototype's material properties. Since model tests are usually performed in water, $\rho = \rho'$. The density, ρ, however, being a representative quantity, stands not only for the density of water but also for the density of the ship's structure. The ship model was therefore to be constructed from a material of the density of steel but with a modulus, E', vastly smaller than that of steel ($E/E' = \ell/\ell'$). Such a material did not exist. It was therefore decided to restrict the model study to bending vibrations. Then analytical knowledge could be infused by expressing Hooke's law as follows:

From $\quad M = EI\,\dfrac{\partial^2 y}{\partial \ell^2} \quad\longrightarrow\quad F \triangleq \dfrac{EI}{\ell^2} \quad\longrightarrow\quad \overline{\pi}_e = \dfrac{F\ell^2}{EI}$

where M is the bending moment, and I is the cross sectional moment of inertia. The Cauchy number would assume the special form of

$$\underline{\overline{Ca}} = \dfrac{\overline{\pi}_e}{Ne} = \dfrac{\ell^4_\rho\,v^2}{EI}$$

With this new Cauchy number, the length scale factor could be selected independent of material properties. For the same material for model and prototype, for instance,

$$\overline{Ca} \quad\longrightarrow\quad \left(\dfrac{v}{v'}\right)^2 = \dfrac{I}{I'}\left(\dfrac{\ell'}{\ell}\right)^4$$

$$\rho = \rho'$$
$$E = E' \quad \text{same material}$$

$$\dfrac{I}{I'} = \left(\dfrac{\ell}{\ell'}\right)^5$$

$$Fr \quad\longrightarrow\quad \left(\dfrac{v}{v'}\right)^2 = \dfrac{\ell}{\ell'}$$

The selected length scale factor was 136. Therefore, a model had to be built,
of the same material as the prototype's, whose area moment of inertia was
136^5 times smaller than that of the prototype. Such a model was built by
composing the hull of nine segments, all joined by a continuous beam to allow
flexures. The area moment, I, of the prototype was calculated in a separate
computer study. The model's beam was varied to make it conform to the model
rule of $I' = I/136^5$, as indicated in Fig. 28. This model was then used to
study the ship's vibratory response at regular and irregular seas.

Use of the Same Material

Up to this point, we have not directly addressed the question of scaling
material properties. It is a cumbersome task. Nature rarely provides the
materials with the properties needed for scale modeling, and to synthesize
materials having specified properties is not easy.

In most cases, attempts to scale material properties end in frustration.
The isotropic Hookean solid, a seemingly easy material to scale, cannot be
scaled at all if both its constitutive properties, Young's modulus, E, and
Poisson's ratio, ν, must be observed, as shown below.

For isotropic, homogeneous, energy-conservative materials, Hooke's law
results in the two representative relations of $\epsilon \triangleq \sigma/E$ and $\epsilon \triangleq \nu\sigma/E$, derived
in the section on "Principal and Common Pi-Numbers." It is obvious that
geometrical similarity between model and prototype (i.e., $\epsilon = \epsilon'$) can be
satisfied only if either the same material is used so that $E = E'$ and $\nu = \nu'$;
or different materials are used with different moduli of elasticity but still
the *same* Poisson's ratio so that $\sigma/E = \sigma'/E'$ and $\nu = \nu'$. These requirements
again suggest use of the same material for model and prototype.

The condition of "same material" can be relaxed only if transverse deformations are disregarded. Then the influence of Poisson's ratio becomes unsub-
stantial, a change that leaves only the requirement of $\sigma/E = \sigma'/E'$ and permits
the use of different materials for model and prototype. The disregard of
Poisson's ratio is an important relaxation in almost all structural problems.

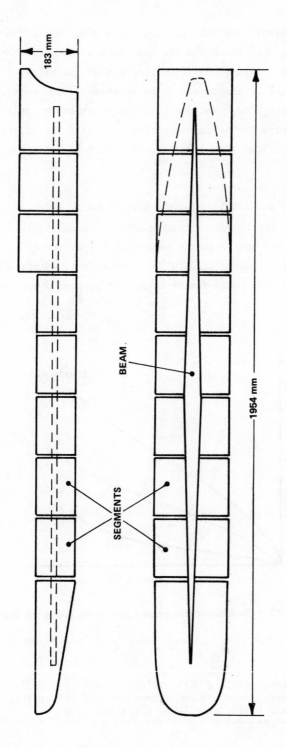

Fig. 28.   Segmented model of aircraft carrier.

Most material constants are not real constants but complex functions of many variables. Hence, even if we use the same material in both the model and the prototype, we cannot ensure identical response of the material unless all participating variables in the constitutive relation of the material have the scale factor of unity. We need not know the constitutive relationship of the material in terms of the participating variables, as pointed out by W.G. Soper,[1] but we must know in advance the list of the constitutive variables.

For instance, if a constitutive relation is known to exist among stress σ, strain ϵ, and strain rate $\dot{\epsilon}$, as indicated schematically in Fig. 29, then we are concerned only with the scale factors of stress and time, and these must be kept at unity; i.e., $\sigma^* = 1$ and $t^* = 1$. Since $\epsilon^* = 1$ (by geometrical similarity), the scale factor of strain rate also becomes unity,

$$\dot{\epsilon}^* = \frac{\epsilon^*}{t^*} = 1$$

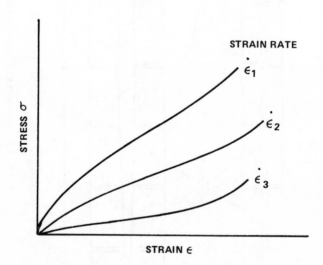

Fig. 29.   Constitutive relation of stress, strain, and strain rate.

[1] W.G. Soper, "Dynamic modeling with similar materials," *Colloquium on Use of Models and Scaling in Shock and Vibration*, W.E. Baker (ed), ASME Winter Annual Meeting (Nov 1963), pp. 51-56, ASME, New York.

In other words, whatever the relation among stress, strain, and strain rate, if we keep the scale factors of stress and time at unity, we know that the constitutive relation $\sigma = \sigma(\epsilon, \dot{\epsilon})$ will be duplicated in the model. Of course, keeping the scale factors of stress and time at unity severely restricts the choice of the scale factors of related quantities such as pressure, force, frequency, speed, and acceleration because, then, $p^* = 1$, $F^* = \ell^{*2}$, $f^* = 1$, $v^* = \ell^*$, $a^* = \ell^*$. If these limitations prove too restricting, further relaxations must be sought. Applications are shown in the case studies entitled "Dynamic Response of Structures," and "Soil-machine Interactions."

SYNOPSIS

- The designer of engineering hardware, in need of experimental data,
  frequently turns to scale modeling, for scale modeling permits him to
  study systems that are usually not tractable in their original form.
  Scale modeling also shortens experimentation by compacting the number
  of variables, and it promotes a deeper understanding of the system under
  investigation.

- Scale model experiments are usually faster and cost less than full-scale
  studies. Yet, if properly designed and executed, they yield sufficient
  information for correct engineering decisions.

- A scale model can serve as a valid substitute for the full-scale system,
  or prototype, because all functional relations of the scale model are
  made similar to corresponding relations of the prototype.

- The scaled transformation of all important prototype relations into similar model relations involves application of five (or less, if less suffice)
  primary scale factors -- those of length, time, force, temperature, and
  electric current.

- By scaling lengths, times, forces, temperatures, and electric currents,
  all quantities (variables and constants) of the model are scaled -- all
  speeds by a speed scale factor, all energies by an energy scale factor,
  all elastic moduli by the elastic modulus scale factor, etc.

- Primary scale factors and their interrelations are derived from the
  fundamental requirement of scale modeling: model and prototype must be
  governed by the same physical laws. Hence, before a model can be designed, the physical laws governing the prototype must be known or, at
  least, hypothesized.

- Each governing law is converted into one (occasionally, more than one) power group of representative quantities. A representative quantity stands for all like quantities. For instance, a representative speed stands for all speeds occurring in a given system -- peak velocity, average velocity, velocity difference, volumetric flow rate per unit area, etc.

- From each power group, a principal pi-number is derived. A principal pi-number is a physical law expressed in terms of representative quantities.

- From correctly identified principal pi-numbers, relations among primary or secondary scale factors (model rules) are derived which, when impressed onto the model, submit it to the laws governing the prototype.

- Principal pi-numbers can be combined in any multiplicative fashion. Principal pi-numbers and combinations thereof often carry the names of renowned physicists and engineers.

- Pi-numbers derived from physical relations other than physical laws (for instance, from $v = \ell/t$) are called common pi-numbers. They are taken for granted and need be spelled out only when needed.

- Thus, the sequence of deriving model design rules is: listing all physical laws governing the prototype to be modeled; converting each law into a relation among representative quantities, i.e., into a principal pi-number (a few laws can be converted into more than one pi-number); extracting from each principal pi-number a relation among primary scale factors.

- If more than one principal pi-number is involved in a scaling problem, conflicting claims on primary scale factors may arise because material properties cannot be arbitrary scaled.

- Conflicting claims on primary scale factors can often be resolved by a skillful lessening of scaling requirements without sacrificing essential information. In this book, more than a dozen relaxation techniques are described.

- To the engineer, scale modeling is a tool, not an end in itself. It
  should be advanced not further than required by the practical task at
  hand. The following case studies demonstrate this important point.

BIBLIOGRAPHY

1. Books and Reports on Dimensional Analysis and Similitude Theory

P.W. Bridgman, *Dimensional Analysis*, Yale University Press, New Haven, 1922 (second printing of revised edition, 1937)

F.W. Lanchester, *Theory of Dimensions and Its Application for Engineers*, Crossby, Lockwood and Sons, London, 1936.

R. Esnault-Pelterie, *L'analyse Dimensionelle* (Dimensional Analysis), Editions F. Rouge and Cie S.A., Lausanne, Switzerland, 1945.

A.W. Porter, *The Method of Dimensions*, Methuen, London, 1946.

G. Murphy, *Similitude in Engineering*, Ronald Press, New York, 1950.

H. Langhaar, *Dimensional Analysis and Theory of Models*, Wiley, New York, 1951, (eighth printing 1967).

H.E. Huntley, *Dimensional Analysis*, Rinehart, New York, 1951.

W.J. Duncan, *Physical Similarity and Dimensional Analysis*, Edwald Arnold, London, 1953.

C.M. Focken, *Dimensional Methods and Their Applications*, Edwald Arnold, London, 1953.

D.C. Ipsen, *Units, Dimensions, and Dimensionless Numbers*, McGraw-Hill, New York, 1960.

E.W. Jupp, *An Introduction to Dimensional Method*, Cleaver-Hume Press, London, 1962.

J. Palacios, *Dimensional Analysis*, MacMillan, London, 1964.

R.C. Pankhurst, *Dimensional Analysis and Scale Factors*, Reinhold, New York, 1964.

A.A. Gukhman, *Introduction to the Theory of Similarity*, Academic Press, New York, 1965.

S.J. Kline, *Similitude and Approximation Theory*, McGraw-Hill, New York, 1965.

P. LeCorbeiller, *Dimensional Analysis*, Appleton-Century-Crofts, New York, 1966.

V.J. Skoglund, *Similitude - Theory and Applications*, International
Textbook, Scranton, Pa, 1967.

H.F. Mullikin, *Dimensional Analysis Handbook, Arrangement Relationships
for Dimensional Analysis (ARDA)*, NASA CR-61634, George C. Marshall
Space Flight Ctr., Huntsville, Ala, 1967.

J.F. Douglas, *An Introduction to Dimensional Analysis for Engineers*, Sir
Isaac Pitman, London, 1969.

J. Pawlowski, *Die Aehnlichkeitstheorie in der physikalisch-technischen
Forschung* (Theory of Similarity in Physical Engineering Research),
Springer Verlag, Berlin, 1971.

N.S. Land, *A Compilation of Nondimensional Numbers*, U.S. Govnt. Printing
Office, NASA SP-274, Washington, DC, 1972.

T.H. Gaiwan, *Dimensional Analysis and the Concept of Natural Units in
Engineering*, U.S. Navy, Naval Postgraduate School, Monterey, Calif,
Mar 1972.

B.S. Massey, *Dimensional Analysis and Physical Similarity*, Van Nostrant
Reinhold, London, 1971   (Actually published in 1973).

T.H. Gaiwan, *On the Axiomatic Foundations of Dimensional Analysis*, U.S.
Navy, Naval Postgraduate School, NPS-57GN74051, Monterey, Calif,
May 1974.

Ballistic Res. Labs., *A Computer Solution of the Buckingham Pi-Theorem
Using SYMBOLANG, a Symbolic Manipulation Language*, M.A. Hirschberg,
BRL-1824, Final Rep., Aberdeen Proving Ground, Md, Aug 1975.

H. Görtler, *Dimensionsanalyse: Theory der physikalischen Dimensionen mit
Anwendungen* (Dimensional Analysis: Theory of Physical Dimensions and
Applications), Springer Verlag, Berlin, 1975.

2.   Books and Reports on Special Fields of Physical Modeling

J. Allen, *Scale Models in Hydraulic Engineering*, Longmans, Green; London,
1947.

R. Esnault-Pelterie, *Dimensional Analysis and Metrology (The Giorgi
System)*, Editions F. Rouge and Cie, S.A., Lausanne, Switzerland, 1950.

R.R. Long, ed., *Fluid Models in Geophysics*, Proc. First Symp. on the Use
of Models in Geophysical Fluid Dynamics, held at Johns Hopkins Univ.,
U.S. Govt. Printing Office, Washington, DC, Sept 1953.

W. Matz, *Anwendung des Aehnlichkeitsgrundsatzes in der Verfahrenstechnik*
(Application of the Principle of Similitude in Process Engineering),
Springer Verlag, Berlin, 1954.

Battelle Memorial Institute, *Similarities in Combustion*, A.E. Weller,
    R.E. Thomas, and B.A. Landry, TR-15038-1, Columbus, O, June 1954.

R.E. Johnstone, M.W. Thring, *Pilot Plants, Models, and Scale-up Methods
    in Chemical Engineering*, McGraw-Hill, New York, 1957.

J.M. Pirie, ed., *Scaling-up of Chemical Plant and Processes*, Joint
    Symposium held at Church House, London, S.W. 1, May 28-29, 1957, The
    Institution of Chemical Engineers, London, 1957.

L.I. Sedov, *Similarity and Dimensional Methods in Mechanics*, Academic
    Press, New York, 1959.

G. Birkhoff, *Hydrodynamics -- A Study in Logic, Fact, and Similitude*
    Princeton Univ. Press, 1960.

W.E. Baker, ed., *Colloquium on the Use of Models in Shock and Vibrations*,
    ASME Winter Annual Meeting, ASME, New York, Nov 1963.

Anon (ed.), *Proc. Symp. on Aeroelasticity and Dynamic Modeling
    Technology*, Sept 23-25, 1963, Dayton, Ohio, Air Force, Research
    and Technology Div. RTD-TDR-63-4197, Pt. I, Mar 1964 (24 papers).

Anon (ed.), *Proc. Conf. on Thermal Scale Modeling*, NASA TM X-51873,
    Washington, DC, Feb 1964 (5 papers).

Virginia Polytechnic Inst., *Proc. Conf. the Role of Simulation in Space
    Technology*, Eng. Extension Series Circulation No. 4 (in 4 parts),
    Aug 1965.

Army Engineer Waterways Experiment Station, *Prototype Performance and
    Model-Prototype Relationship*, F.B. Campbell and E.B. Pickett, Misc.
    Paper No. 2-857, Vicksburg, Miss, Nov 1966 (Hydraulic models).

T.M. Charlton, *Model Analysis of Plane Structures*, Pergamon Press, New
    York, 1966.

J.P. Comstock, ed., *Principles of Naval Architecture*, Soc. of Naval
    Architects and Marine Engineers, New York, 1967.

H. Ramberg, *Gravity, Deformation, and the Earth's Crust (as Studied by
    Centrifuged Models)*, Academic Press, New York, 1967.

H.J. Cowan, J.S. Gero, G.D. Ding, and R.W. Muncey, *Models in Architecture*,
    Elsevier, New York, 1968.

W.E. Baker, P.S. Westine and F.T. Dodge, *Target Vulnerability Scaling and
    Modeling*, Joint Munitions Effectiveness Manual - (Air to Surface), U.S.
    Force Technical Handbook 61 JTCG/ME-69-1, Jan 1969.

D. Phillips-Birt, *Ship Model Testing*, John DeGraff, Tuckahoe, New York
    1970.

A.N. Barkhatov, *Modeling of Sound Propagation in the Sea*, Transl. from Russian by J.S. Wood, Consultants Bureau, New York, 1971.

J. Zierep, *Similarity Laws and Modeling*, Gasdynamics Series, Vol. 2, Marcel Dekker, New York, 1971.

P.R.H. Verbeek, *Schaalregels voor Pompen* (Scaling Rules for Pumps), Waaterloopkundig Laboratorium, Delft, Netherlands, Publ-96, Nov 1971.

B. Hilson, *Basic Structural Behavior Via Models*, Wiley, New York, 1972.

Anon. (ed)., *Proc. Symp. on Status of Testing and Modeling Techniques for V/STOL Aircraft*, 26-28 Oct 1972, Essington, Pa, Am.Helicopter Soc., (Many papers on aerodynamic and acoustic modeling techniques).

R. Kurth, *Dimensional Analysis and Group Theory in Astrophysics*, Pergamon Press, New York, 1972.

J.G. Waugh and G.W. Stubstaad, *Hydroballistics Modeling*, U.S. Government Printing Office, Washington, DC, 1973.

W.E. Baker, P.S. Westine, and F.T. Dodge, *Similarity Methods in Engineering Dynamics*, Hayden, Spartan Books, Rochelle Park, N.J., 1973.

Air Force Weapons Lab., *An Evaluation of Dimensional Analysis and Modeling Techniques for Air Force Civil Engineering Research*, J.F. Kanipe, AFWL-TR-72-179, Kirtland AFB, N. Mex, Mar 1973 (289 references on modeling dynamic responses of structures, soil, and pavements).

H. Hossdorf, *Model Analysis of Structures* (Transl. by C. VanAmerongen), Van Nostrand Reinhold, New York, 1974.

3. Chapters in Books and Reports

J.E. Warnock,"Hydraulic similitude," Ch. 11, *Engineering Hydraulics*, Proc. Fourth Hydraulics Conf., Wiley, New York, 1950.

J.B. Wilbur and C.H. Norris, "Structural model analysis," Ch. 15, *Handbook of Experimental Stress Analysis*, M. Hetenyi (ed.), Wiley, New York, 1950.

J.N. Goodier, "Dimensional analysis," Appendix II, *Handbook of Experimental Stress Analysis*, M. Hetenyi (ed.), Wiley, New York, 1950.

D.E. Hudson, "Scale-model principles," Ch. 27, *Shock and Vibration Handbook*, C.M. Harris and C.E. Crede (eds.), Vol. 2, McGraw-Hill, New York, 1961.

W.H. McAdams, "Dimensional analysis," Ch. 5, *Heat Transmission*, Mc-Graw-Hill, New York, 1954.

R.L. Bisplinghoff, H. Ashley, and R.L. Halfman, "Aeroelastic model theory,"
"Model design and construction," Chs. 11 and 12, *Aeroelasticity*,
Addison-Wesley, Cambridge, Mass, 1955.

D.G. Shepard, "Dimensional analysis," Ch. 2, *Principles of Turbomachinery*,
Macmillan, New York, 1956.

A.B. Metzner and R.L. Pigford, "Scale-up theory and its limitations,"
Ch. 2, *Scale-up in Practice*, R. Fleming (ed.), Reinhold, New York,
1958.

M. Holt, "Dimensional analysis," Ch. 15, *Handbook of Fluid Dynamics*,
V.L. Streeter (ed.), McGraw-Hill, New York, 1961.

E. Weber, "Dimensional analysis," Ch. 3, *Handbook of Engineering Fundamentals*, O.W. Eshbach (ed.), Wiley, New York, 1961.

W.G. Molyneux, "A consideration of the similarity requirements for aerothermoelastic tests on reduced scale models," pp. 290-339, *Proc. Symp.
on Aerothermoelasticity*, USAF, Aeronautical Syst. Div., TR-61-645,
1961.

W.S. von Arx, "Laboratory models," Ch. 10, *An Introduction to Physical
Oceanography*, Addison-Wesley, Reading, Mass, 1962.

D.H. Norrie, "Dimensional analysis and theory of models," Ch. 3., *An
Introduction to Incompressible Flow Machines*, Elsevier, New York, 1963.

G.F. Wislicenus, "Similarity considerations," Part II, *Fluid Mechanics of
Turbomachinery*, Dover, New York, 1965.

W. Rocha, "Structural model techniques," Ch. 16, *Stress Analysis*, O.C.
Zienkiewicz and G.S. Holister (eds.), Wiley, New York, 1965.

G.H. Keulegan, "Model laws for coastal and esturine models," Ch. 17,
Estuary and Coastline Hydrodynamics, A.T. Ippen (ed.), McGraw-Hill,
New York, 1966.

H.B. Simmons, "Tidal and salinity model practice," Ch. 18, *Estuary and
Coastline Hydrodynamics*, A.T. Ippen (ed.), McGraw-Hill, New York, 1966.

C.A.M. King, "Experiment and theory - models," Ch. 4, *Techniques in
Geomorphology*, St. Martin's Press, New York, 1966.

M.S. Morgan, "Hardware models in geography," Ch. 17, *Models in Geography*,
R.J. Chorley and P. Haggett (eds.), Methuen, London, 1967.

A.J. Stepanoff, "Model testing and scale factors," Ch. 10, *Gravity Flow
of Bulk Solids and Transportation of Solids in Suspension*, Wiley,
New York, 1969.

B. Günther, "Stoffwechsel und Koerpergroesse -- Dimensionsanalyse und
   Similaritaetstheorien" (Metabolism and body size -- dimensional
   analysis and similarity theories), in *Physiologie des Menschen*,
   Vol.II (Energiehaushalt und Temperaturregulation) pp. 117-151. Urban
   und Schwarzenberg, Munich, 1971.

F.C. Frischknecht, "Electromagnetic scale modeling," Ch. 8, *Electromagnetic
   Probing in Geophysics*, J.R. Wait (ed.), The Golem Press, Boulder,
   Colo, 1971.

J. Bogardi, "Hydraulic similarity in sediment transport," Ch. 2.5,
   Sediment Transport in Alluvial Streams (Transl. by Z. Szilvassy),
   Akademiai Kiado, Budapest, 1974.

4. Fundamental Articles and Papers on Theory and Applications of Physical
    Modeling

E. Buckingham, "On physically similar systems; illustrations of the use of
   dimensional equations,"*Phys. Rev.* 4, 2, 345-376 (1914).

J.W.S. Rayleigh, "The principle of similitude," *Nature* 95, 2368, 66-68
   (Mar 1915); and 95, 2389, 614 (Aug 1915).

E. Buckingham, "The principle of similitude," *Nature* 96, 2406, 396-397
   (Dec 1915).

M. Weber, "Die Grundlagen der Aehnlichkeitsmechanik and ihre Verwertung
   bei Modellversuchen," (The principles of physical similarity and their
   use in model testing), *Jahrbuch der Schiffbautechnischen Gesellschaft*
   20, 355-477 (1919).

M. Weber, "Das Allgemeine Aehnlichkeitsprinzip der Physik und sein
   Zusammenhang mit der Dimensionslehre und der Modellwissenschaft,"
   Jahrbuch der Schiffsbautechnischen Gesellschaft 31, 274-388 (1930)
   (David Taylor Model Basin Transl. 300 under the title, "The universal
   principle of similitude in physics and its relation to the dimensional
   theory and the science of models").

M.K. Hubbert, "Theory of scale models as applied to the study of geologic
   structures," *Bull. Geol. Soc. Amer.* 48, 1459-1520 (Oct 1, 1937).

E.R. van Driest, "On dimensional analysis and the presentation of data in
   fluid flow problems," *J. Appl. Mech.* 13, 1, A-34-40 (Mar 1946) (ASME
   Transl. 68).

A. Klinkenberg and H.H. Mooy, "Dimensionless groups in fluid friction,
   heat, and material transfer," *Chem. Eng. Progress* 44, 1, 17-36 (Jan
   1948).

G. Sinclair, "Theory of models of electromagnetic systems," *Proc. Inst.
   Radio Engrs.* 36, 11, 1364-1370 (Nov 1948).

F.H. Todd, "The fundamentals of ship model testing," *Soc. Naval Architects and Marine Engs., Trans.* 59, 850-896 (1951).

H. Rouse, "Model techniques in meteorological research," *Compendium of Meteorology*, Amer. Meteorol. Soc., Boston, Mass (1951).

P. Franke, "Similitude requirements in hydraulic models" (in German), *Mitt. Versuch. Wasserbau der Techn. Hochschule München* No. 1, 80-100 (1953).

F. Schultz-Grunow, "Neue Anwendungen der Aehnlichkeitstheorie" (New applications of the theory of similarity), *Chemie-Ingenieurwesen-Technik* 26, 1, 18-24 (Jan 1954).

T.H. Higgens, "Electroanalogic methods - Part IV," *Applied Mechanics Reviews* 10, 8, 331-335 (Aug 1957) (more than 200 references on dimensional analysis, similitude, theory of models, and scaling).

D.F. Boucher and G.E. Alves, "Dimensionless numbers for fluid mechanics, heat transfer, and chemical reactions," *Chem. Eng. Progress* 55, 9, 55-64 (Sep 1959).

J. Dugundji and J.M. Calligeros, "Similarity laws for aerothermoelastic testing," *J. Aerospace Sciences* 29, 8, 935-950 (Aug 1962).

W.R. Stahl, "The analysis of biological similarity," *Advances in Biological and Medical Physics* 9, 355-464 (1963).

J. Hayes, "Structural modeling," California Inst. Tech., Jet Propulsion Laboratory, Literature Search No. 523 (119 references on aeroelastic model tests from 1927 to 1963) (Mar 1963).

F.V.A. Engel, "Das vollstaendige System dimensionsloser Kenngroessen der Aehnlichkeitsgesetze fuer Stroemungsvorgaenge und Waermeuebertragung" (The complete system of dimensionless groups and model laws in fluid dynamics and heat transfer), *VDI Zeits.* 107, 15, 671-676 (May 1965); and 107, 18, 793-797 (June 1965).

W.W. Happ, "Computer-oriented procedures for dimensional analysis," *J. Appl. Phys.* 38, 10, 3918-3926 (Sep 1967).

R.N. White, "Similitude requirements for structural models," Paper 469, presented at ASCE Structural Eng. Conf., Seattle Wash, May 1967.

L. Borel, "Characteristic dimensionless figures for turbo-machines," *Water Power* 19, 12, 494-498 (Dec 1967); and 20, 1, 27-32 (Jan 1968).

A.D. Sloan and W.W. Happ, "Computer program for dimensional analysis," NASA TN D-5165, Washington, DC, Apr 1969.

R.L. Causey, "Derived measurement, dimensions, and dimensional analysis," *Philosophy of Science* 36, 3, 252-270 (Sept 1969).

D.F. Young, "Basic principles and concepts of model analysis," *Experimental Mechanics* 11, 7, 325-336 (July 1971).

R.P. Kroon, "Dimensions," *J. The Franklin Institute* 292, 1, 45-55 (July 1971).

E.O. Macagno, "Historico-critical review of dimensional analysis," *J. The Franklin Institute* 202, 6, 391-402 (Dec 1971).

G. Murphy, "Models with incomplete correspondence with the prototype," *J. The Franklin Institute* 292, 6, 513-518 (Dec 1971).

R.L. Shannon, "Thermal scale modeling of radiation-conduction-convection systems," *J. Spacecraft and Rockets* 10, 8, 485-492 (Aug 1973).

S.Swarup and R.K. Tewari, "Scale-model techniques for radio wave propagation studies," *Indian J. of Radio and Space Physics* 3, 21-27 (Mar 1974).

H. Görtler, "Zur Geschichte des Pi-Theorems" (History of the Pi-Theorem), *Zeitschrift fuer angewandte Mathematik und Mechanik* 55, 3-8 (Jan 1975).

J.J. Sharp, "Application of dimensional reasoning to thermal systems," *J. The Franklin Institute* 299, 3, 191-197 (Mar 1975).

R.L. Bannister, "Structural models for vibration control," *Noise Control Engineering* 4, 2, 84-92 (Mar-Apr 1975).

PART II

APPLICATIONS

The following case studies are gleaned from the open literature and, to a much lesser degree, abstracted from the author's own work.

Scale modeling can be applied to virtually all scientific and engineering fields. For several years, the author has been scanning reviews and abstracts, together with hundreds of candidate articles and papers. Most of these were eventually discarded for lack of originality, for deficiency in experimental evidence or analytical support, or because they duplicated other case studies. From the remaining hundred cases or so, a final selection of less than a dozen had to be made -- a task inevitably involving some subjective judgments. The author hopes, nevertheless, to have assembled a collection of case studies both instructive and representative of modern scale model work.

All selected case studies have been reworked into a uniform format. The original similitude analyses were revised and adapted to the law approach, the method advocated in this book. Some descriptions of experimental apparatus have been curtailed, some numerical results have been reinterpreted, and some diagrams have been redrawn; the responsibility for these changes rests with the author, of course. As a result, all case studies are uniformly presented, with background information given before a statement of the problem. Next is a physical interpretation of the stated problem to prepare for the subsequent development of governing laws, pi-numbers, and model rules. A discussion of conflicting model requirements and their relaxation precedes descriptions of experiments and results. Finally, the source of information is given, together with some related publications.

Dynamic Response of Structures

Rapid deformations of solid bodies and structures take place everywhere --
in forging mills, at car accidents, on battlefields, during earthquakes,
and on tennis courts. Yet, because of the complex processes of energy
interchange between the interacting media, no satisfactory analytical solutions have come forth. On the other hand, experimentation can be cumbersome
or expensive, particularly if the involved structures are large. As a
result, deformation processes are frequently studied with the help of scale
models.

Viewed in the most general way, dynamic deformation of solid structures and
bodies is governed by at least two forces: inertial forces of the rapidly
displaced masses, and stress forces of the deformed material. In representative terms,

Newton's law
of inertia

$$F \triangleq ma \longrightarrow F \triangleq \rho \frac{\ell^4}{t^2} \longrightarrow \pi_i = \frac{F t^2}{\rho \ell^4}$$

$$m \triangleq \rho \ell^3 \qquad a \triangleq \frac{\ell}{t^2}$$

Definition of
stress force

$$F \triangleq \sigma \ell^2 \longrightarrow \pi_\sigma = \frac{F}{\sigma \ell^2}$$

Eliminating the representative force, we obtain

$$\pi_1 = \frac{\pi_\sigma}{\pi_i} = \frac{\rho \ell^2}{\sigma t^2}$$

Density, length, and time can be characterized in unequivocal terms for each
given case. Stress, σ, by contrast, is a function of strain and, for many
materials, of time as well. If this function were known, it could be
formulated as a law and, hence, expressed as a principal pi-number. For

deformations beyond the elastic state, however, no such function is available
for many materials. Under these circumstances, the same material is used
for model and prototype, with the following consequences.

By definition, homologous behavior requires equal strains of all corresponding elements of model and prototype, at corresponding times. Using the
same material would indeed impose equal strains if the stress scale factor
and the time scale factor were kept unity so that corresponding stresses and
corresponding times would be equal. Consequently, corresponding strains
would be equal, too. Unfortunately, with $\sigma = \sigma'$ and $t = t'$, and $\rho = \rho'$ as
well (since the same material is used), the length scale factor would be
reduced to unity, and modeling would be impossible.

$$\pi_1 = \frac{\rho \ell^2}{\sigma t^2} \quad\quad\quad \begin{array}{l} \sigma = \sigma' \\ \rho = \rho' \\ t = t' \end{array} \Big\} \text{ same material} \quad \longrightarrow \ell = \ell' \longrightarrow \text{ no modeling}$$

In many cases, strain rate effects can be ignored. Then the time scale
factor can be freely chosen, and modeling becomes possible, as in the following two problems.

1a. Penetration of Projectiles

The problem was to study, with the help of models, the penetration of cone-
nosed steel projectiles into 5083 aluminum armor. The model projectiles,
0.50 caliber bullets, had a total cone angle of 35 deg.

Besides the neglection of rate effects and the use of the same material, the
following assumptions were made.

The target was viewed as supported by a semi-infinite body so that it could
not move as a whole. In addition, the deformations of the target were considered restricted to the vicinity of the impact area. Consequently, the
kinetic energy of the target after impact was assumed to be small.

The target is softer than the projectile. Thus, since the projectile was considered to deform very little, most of its kinetic energy would be used to deform the target. Only a negligible part was thought to be transformed into friction heat or carried away by traveling stress waves.

No residual kinetic energy of the projectile could be left since the target was thick enough to defeat the projectile. In fact, tests were limited to penetrations less than the cone height.

As a consequence of these relaxations, only two kinds of energies were modeled -- the kinetic energy of the impacting projectile, and the deformation energy of the armor.

Kinetic energy
(Newton's law
of inertia)
$$E \triangleq m v^2 \longrightarrow \pi_i = \frac{E}{m v^2}$$

Stress energy
$$E \triangleq \sigma \ell^3 \longrightarrow \pi_\sigma = \frac{E}{\sigma \ell^3}$$

In content, these two pi-numbers are identical with the two numbers developed earlier. Eliminating the representative energy, we obtain

$$\pi_2 = \frac{\pi_i}{\pi_\sigma} = \frac{\sigma \ell^3}{m v^2} \longrightarrow \frac{y}{E^{1/3}}$$

same material
(no rate effects) $\sigma = \sigma'$ $\quad \ell \triangleq y$

$m v^2$ interpreted as kinetic energy
of bullet before impact, E

where y is the penetration depth of the projectile. Since penetrations were limited to less than cone height, all indentations were geometrically similar to each other. Therefore, the same projectile could be used as model and prototype as long as the penetration did not surpass the cone length. Since the armor was thought to be infinitely large, it too could be treated as model and prototype.

Identical 0.50 caliber projectiles were driven into heavy blocks of 5083
aluminum. For the model tests, the projectiles were dropped at various
heights from a drop tower; for the prototype test, they were fired by charges
of various explosive energies. Figure 1.1 shows the measured ratios of $y/E^{1/3}$
plotted versus the impact speed, v. In the absence of strain rate effects,
the ratios should be independent of speed. As the figure demonstrates, this
expectation is nearly satisfied; although the impact velocities vary widely,
the ratio stays nearly constant. Hence, the assumption of rate effects
being negligible was justified.

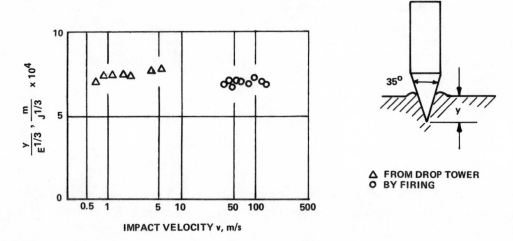

Fig. 1.1   Cone penetration of aluminum.

Source of Information

W.G. Soper, "Dynamic modeling with similar materials," *Colloquium on Use of
Models and Scaling in Shock and Vibration*, W.E. Baker (ed), ASME Winter
Annual Meeting pp. 51-56, ASME, New York (Nov 1963).

1b. Explosive Forming of Sheet Metal

Metal sheets can be formed by using an explosive charge to force a blank
into a suitable die, as illustrated in Fig. 1.2. Metal-forming rates may
reach several hundred meters per second, in contrast to conventional drawing
and extruding operations normally conducted at rates of only 0.5 to 1.5 m/s.
Developing an optimum technique is a trial-and-error process, one that may
be carried on most conveniently on a small scale at a relatively low cost.

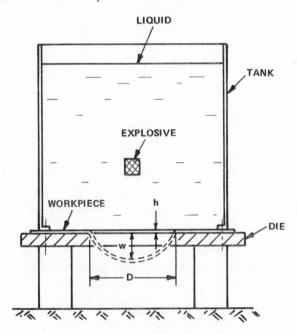

Fig. 1.2.   Forming die and workpiece.

The energy of the explosive is transmitted through the fluid in the tank
to the metal blank. The process is governed by the total explosive energy,
the inertial effects of both the fluid and the blank, and the stress energy
due to deformation of the blank. The (practically infinitely) large mass
of the die need not necessarily be scaled, and viscous effects of the fluid
can be safely neglected. To alleviate the difficulties of scaling material
properties, the same fluid and the same blank material were used in the model
as in the prototype. By doing so, the representative stress was made the
same, and all corresponding stresses, including the hold-down pressures (of

the blank) provided by explosive energy, were rendered equal. Rate effects
were avoided by choosing a material for the blank whose stress would be not
affected by the strain rate.

With these assumptions, the earlier developed pi-numbers apply.

$$\pi_1 = \frac{\rho \ell^2}{\sigma t^2} \quad\xrightarrow{\hspace{4cm}}\quad v = v'$$

$$\rho = \rho'$$
$$\sigma = \sigma'$$

same material (no rate effects)

$$\pi_\sigma = \frac{E}{\sigma \ell^3} \quad\xrightarrow{\hspace{4cm}}\quad \frac{E}{E'} = \left(\frac{\ell}{\ell'}\right)^3$$

The identity of π_σ calls for the total explosive energy, E, to be made
proportional to ℓ^3; and the identity of π_1 calls for the shock waves released
by the explosions to travel at the same corresponding speeds in both model
and prototype fluids. This requirement was assumed to be satisfied by using
the same fluids.

Tests were performed with geometrically similar forming dies, from 12.2 to
61 cm in diameter, and with scaled thicknesses of metal blank, draw radii of
the die, tank diameter, blank sizes, diameters of charge, depths of water,
and standoff distances of the explosive charges. The test materials consisted of metal blanks of 2014 aluminum alloy, whose stress would hardly be
affected by strain rate.[1] The same type of explosive (25% Powertol No. 7)
and the same water were used for all tests.

Test results were plotted in form of π_σ versus dimensionless maximum blank
deformation, w/D , Fig. 1.3. In π_σ, blank thickness h was taken as representative length; σ is the static yield stress of the metal blank. The
explosive energy was given in grains of the explosive material, as is customary
in U.S. ordnance (1 grain is defined as 1/7000 of a pound mass avoirdupois or
0.0648 g).

---

[1]See Case Study 1a.

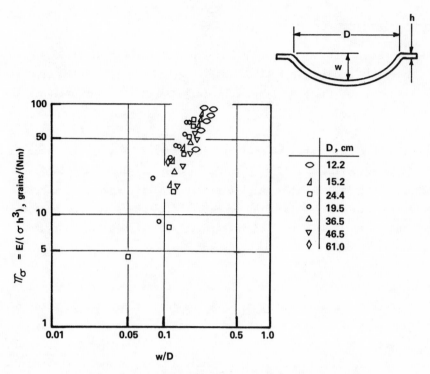

Fig. 1.3.   Explosive energy vs. blank deformation.

The plot convincingly confirmed the model design rules applied.

Sources of Information

A.A. Ezra and F.A. Penning, "Development of scaling laws for explosive forming," *Experimental Mechanics* 2, 8, 234-239 (Aug 1962).

A.A. Erza, "Scaling laws and similitude requirements for valid scale model work," *Colloquium on Use of Models and Scaling in Shock and Vibration*, W.E. Baker (ed), ASME Winter Annual Meeting, 57-63, ASME, New York (Nov 1963).

1c.   Blast-Loaded Structure

In recent years, efforts have been made to remove the restrictions of the
"same material" modeling practices and to use models fabricated from material
other than that of the prototype. The following case study presents an
example.

One of the most important tasks of the designer of military hardware is to
predict the vulnerability of structures such as airplanes and land vehicles
to air blasts from detonating high-explosive shells, warheads, mines, etc.
Since full-scale testing is often prohibitively expensive, the potential
of homologous modeling with different materials has been investigated,
using simple structures such as beams and cylinders.

Choosing a different material for the model offers greater freedom in model
design. Physical similarity requires a constant stress scale factor over
the whole range of stress application, a requirement that is easily satisfied if model and prototype follow Hooke's law. Then the stress scale
factor is $\sigma^* = E_y / E_y'$ (E_y is Young's modulus). If, however, the material
is non-Hookean or if it is deformed beyond the yield point, the condition
of constant stress scale factor requires matching the stress-strain *curves*
in such a way that the model curve multiplied by the stress scale factor
duplicates the prototype curve.

Figure 1.4 depicts two stress-strain curves of a prototype and a model
material -- heat-treated Inconel X, a high-yield iron-nickel-cobalt alloy
for the prototype; and 6061-T6 aluminum alloy for the model. The model
curve approximately duplicates the prototype curve if multiplied by the
stress scale factor of $\sigma^* = 3.1$.

As a consequence of employing different materials for model and prototype,
the materials' density must also be scaled, in accordance with the previously
derived pi-numbers and model design rules.

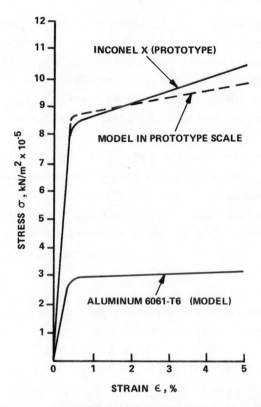

Fig. 1.4.   Stress-strain curves of model and prototype.

STRESS SCALE FACTOR $\sigma^* = 3.1$

Kinetic energy $E \triangleq ma\,l$ $\xrightarrow{\qquad m \triangleq \rho l^3 \qquad a \triangleq \frac{l}{t^2} \qquad t \triangleq \frac{l}{v} \qquad}$ $E \triangleq \rho v^2 l^3$

Stress energy $E \triangleq \sigma l^3$ $\xrightarrow{\hspace{6cm}}$

$\sigma \triangleq \rho a_s^2 \xleftarrow{\qquad v \triangleq a_s \qquad} \sigma \triangleq \rho v^2$

$a_s = a_s'$

$\dfrac{\rho}{\rho'} = \dfrac{\sigma}{\sigma'}$ or $\rho^* = \sigma^*$

With the representative velocity, v, interpreted as the speed of sound, a_s, and keeping the speed of sound the same for model and prototype, the density scale factor must equal the stress scale factor. The density of the proto-type material was 8450 kg/m^3; that of the model material, 2710 kg/m^3. Thus, the density scale factor of the two materials was $\rho^* = 3.11$ -- nearly the stress scale factor, as required. For blast loading, the density scale factor encompasses not only the density of the structures but also the den-sity of the air since the inertia of the air is what effects the deformations of the structures. Consequently, the model tests of our case study were performed in a spherical enclosure with 1/3 the atmospheric pressure, whereas the prototype tests took place in the open field. Lowering the air pressure and thus the air density (but keeping the temperature the same) would not affect the speed of sound; therefore, the condition of $a_s = a_s'$ remained preserved.

Under these -- fortunate -- circumstances, model tests with different material seemed possible. Various simple structures were tested, among them cantilever beams -- 76.2 mm long, 6.35 mm wide, and 1.59 mm thick for the prototype; and 1.75 times smaller for the model. The beams were exposed in pairs to Pentolite charges as indicated in Fig. 1.5 (the prototype in the open field, the models in the Blast Sphere Facility of the U.S. Army Ballistic Research Laboratory), and their permanent tip deflections, δ, measured. By changing the standoff distance R, various tip deflections were achieved. The explosive charges were scaled according to

$$\frac{E}{E'} = \frac{\sigma}{\sigma'}\left(\frac{\ell}{\ell'}\right)^3 \longrightarrow \frac{E}{E'} = 16.6$$

$$\frac{\sigma}{\sigma'} = 3.1 \qquad \frac{\ell}{\ell'} = 1.75$$

Figure 1.5 shows the results. The correlation between model and prototype results must be considered excellent.

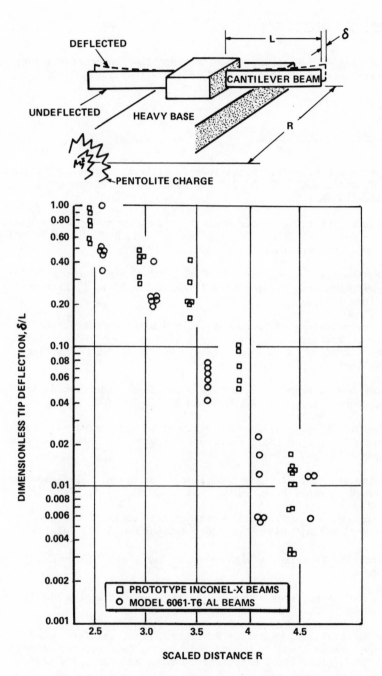

Fig. 1.5.   Permanent tip deflection of cantilever beam.

Source of Information

W.E. Baker and P.S. Westine, "Modeling the blast response of structures using dissimilar materials," *AIAA J.* 7, 5, 951-959 (May 1969).

Selected Related References

W.G. Soper and R.C. Dove, "Similitude in package cushioning," *Trans. ASME, Ser. E, J. Appl. Mech.* 29, 2, 263-266 (Jun 1962). Discussion by G.F. Nevill, 30, 1, 152-153 (Mar 1963).

J. Sperrazza, "Modeling of air blast," *Colloquium on Use of Models and Scaling in Shock and Vibration*, W.E. Baker (ed), ASME Winter Annual Meeting, 65-70, ASME, New York (Nov 1963).

Air Force Special Weapons Ctr., *Similitude Studies of Reentry Vehicle Response to Impulsive Loading*, by G.E. Nevill, AFSWC-TDR 63-1, 1, Kirtland AF Base, N. Mex. (Mar 1963).

W.E. Baker, "Modeling of large elastic and plastic deformations of structures subjected to transient loading," *Colloquium on Use of Models and Scaling in Shock and Vibration*, W.E. Baker (ed), ASME Winter Annual Meeting, 71-78, ASME, New York (Nov 1963).

D.F. Young and G. Murphy, "Dynamic similitude of underground structures," *J. Engr. Mech. Div., Proc. ASCE* 90, EM 3, 111-133 (Jun 1964).

R.K. Tener, "The application of similitude to protective construction research," *Proc. Symp. Soil-Structure Interaction*, Univ. of Arizona, Tuscon, Ariz., Sep 1964, pp. 296-302.

D.R. Denton and W.J. Flathau, "Model study of dynamically loaded arch structures," *J. Engr. Mech. Div., Proc. ASCE* 92, EM 3, 17-32 (Jun 1966).

B.P. Denardo et al., "Projectile size effects on hypervelocity impact craters in aluminum," NASA TN D-4067, Washington, D.C., Jul 1967.

D.I. Rabinovich, "Modeling a high-velocity air blast," *Thermal Eng.* 14, 2, 79-81 (1967).

W.H. Kim and C. Kisslinger, "Model investigations of explosions in prestressed media," *Geophys.* 32, 4, 633-651 (Aug 1967).

U.S. Army Materials and Mechanics Res. Ctr., *Proc. 1st Intl. Conf. Center for High Energy Forming* 1 and 2, Denver Res. Inst., Denver Colo. (1968) [See also Proc. 2nd Conf. (June 1969) Estes Park, Colo.; and Proc. 3rd Conf. (July 1971), Vail, Colo.].

R.S. David, "Explosive working of metals," Ch. 37 *Techniques of Metals 1: Techniques of Materials Preparation and Handling*, Pt. 3, 1589-1607, R.F. Bunshah (ed), Interscience Publ., Wiley, New York (1968).

J.M. Jordaan, Jr., "Simulation of waves by an underwater explosion," *J. Waterways and Harbors Div., Proc. ASCE* 95, WW 3, 355-377 (Aug 1969).

J.M. Horowitz and G.E. Nevill, "A correction technique for structural impact modeling using dissimilar materials," *AIAA J.* 7, 8, 1637-1639 (Aug 1969).

U.S. Army Engr. Waterways Experiment Station, *Dynamic Tests of a Model Flexible-Arch-Type Protective Shelter* by T.E. Kennedy, AEWES - Misc. Paper N-71-3, Vicksburg, Miss., Apr 1971.

Southwest Res. Inst., *Model Analysis of the Response of Armor Plate to Land Mine* by P.S. Westine and W.E. Baker, San Antonio, Tex., Feb 1971.

Aeronautical Res. Associates of Princeton, *A Brief Study of the Possibility of Using Partial Similitude to Estimate the Level of Impact Damage* by C.P. Donaldson and B.H. Jones, ARAP 165, Princeton, N.J., July 1971.

B.L. Morris, "Measurement of impulse from scaled buried explosives," *The Shock and Vibration Bull.* Pt. 2, 123-127 (Jan 1972).

R.L. Peterson and E.O. Roberts, "Experimental investigation and correlation of the ground impact acceleration characteristics of a full scale capsule and a 1/4 scale model aircraft emergency crew escape capsule system," Paper 73-480 presented at 4th AIAA Aerodynamic Deceleration Systems Conf., Palm Springs, Calif, May 1973.

W.E. Baker, P.S. Westine, and F.T. Dodge, *Similarity Methods in Engineering Dynamics, Theory and Practice of Scale Modeling*, Hayden, Rochelle Park, N.J., 1973 (Appr. 250 references on dynamic modeling).

Air Force Weapons Lab., *An Evaluation of Dimensional Analysis and Modeling Techniques for Air Force Civil Engineering Research* by J.F. Kanipe, AFWL-TR-72-179, Kirtland AFB, N. Mex., Mar 1973 (289 references on dynamic modeling).

M. Held, "Das Cranz'sche Modellgesetz" (The Cranz model law), in: Safety Technology: Generation and Action of Explosive Systems; Annual Meeting, Inst. Treib-und Explosivstoffe, Berghausen bei Karlsruhe, 1974.

Ballistic Res. Lab., *Blast Loading of Objects in Basement Shelter Models* by G.A. Coulter, BRL-MR-2348, Aberdeen Proving Ground, Md., Jan 1974.

P.H. Thornton, "Static and dynamic collapse characteristics of scale model corrugated tubular sections," *Trans. ASME* 97, H4, 357-362 (Oct 1975).

Related References on Vehicle Collisions

M.A. MaCaulay and R.G. Redwood, "Small scale model railway coaches under impact," *The Engineer* 218, 5683, 1041-1046 (Dec 1964).

W. Endress and H. Hagen, "Occurrences during and after the head-on collision of automobiles -- investigated on model vehicles," (in German), *Automobiltechnische Zeitschrift* (ATZ), 68, 3, 86-90 (Mar 1966).

Stevens Inst. Technology, Davidson Lab., *Highway Center-Barrier Investigation; Pt. II -- Model Study* by J.A. Starrett and I.R. Ehrlich, TR-1139, Hoboken, N.J., Jun 1967.

D.I. Cook, "Model aid accident reconstruction and analysis," *Traffic Eng.* 37, 6, 34-36 (Mar 1967).

M.P. Jurkat and J.A. Starrett, "Automobile barrier impact studies using scale model vehicles," *Highway Res. Record 174* Natl. Res. Coun., Highway Res. Board., Natl. Acad. Sci., Washington, D.C., 1967, pp. 30-41.

R.I. Emori a·¡ D. Link, "A model study of automobile collisions" (Paper 690070 presented at ᴜAE Intl. Auto. Engr. Congr., Detroit, Mich., Jan 1969).

Wyle Lab., Res. Staff, *A Scale Model Study of Crash Energy Dissipating Vehicle Structures* by G.C. Kao and G.C. Chan, 68-3, Vol. V, Huntsville, Ala., Mar 1968 (See also Defense Dept. Bull. 39, pt. 4, Mar 1969).

A. Morelli, "Collision tests on 1/15 scale model vehicles," Paper 2-05 presented at 12th Intl. Automobile Tech. Congr (FISITA), Barcelona Spain, May 1968. Transl. into English: MIRA Transl. No. 8/69, Lindley, Warwickshire, Engl. (MIRA = Motor Ind. Res. Assoc.).

M.A. Kaplan, R.J. Hensen, and R.J. Fay, "Space technology for auto-highway safety," *Highway Res. Record 306*, Natl. Res. Coun., Highway Res. Board, Natl. Acad. Sci., Washington, D.C., 1970, pp. 25-38.

R.J. Fay and E.P. Wittrock, "Scale model test of an energy-absorbing barrier," *Highway Res. Record 343*, Natl. Res. Coun., Highway Res. Board, Natl. Acad. Sci., Washington, D.C., 1971, pp. 75-82.

W.T. Lowe, S.T.S. Al-Hassani, and W. Johnson, "Impact behavior of small scale model motor coaches," *J. Automotive Eng.* (JAE) 3, 19-26 (Jan 1972).

R.I. Emori, "Scale models of automobile collisions with breakaway obstacles," *Experimental Mechanics* 13, 2, 64-69 (Feb 1973).

T.O. Jones and W.A. Elliot, "An introduction to scale model testing to determine air cushion crash sensor location," *Proc. 3rd Intl. Conf. on Occupant Protection* SAE, New York, 1974, pp. 314-322.

B.S. Holmes and G. Sliter, "Scale modeling of vehicle crashes -- techniques, applicability, and accuracy: cost effectiveness," *Proc. 3rd Intl. Conf. on Occupant Protection* SAE, New York, 1974, pp. 323-348.

J.S. Westcott, "Safety Motorcycle," *Proc. Instn. Mech. Engs.* 189 1/75, 1-16 (1975).

CASE STUDY 2

Cavitation at Missile Entry into Water

A liquid is said to cavitate when bubbles or larger cavities are observed
to form as a consequence of pressure reduction. The bubbles and cavities
are filled with vapor; they can seriously reduce the efficiency (and, when
violently collapsing near a solid boundary, the life) of hydraulic pumps,
turbines, valves, ship propellers, hydrofoils, torque converters and other
hydraulic machinery. Cavitation has therefore received considerable
experimental and analytical attention for many decades, with the first
experimental studies started by Reynolds as early as 1874. Yet, present
theoretical concepts of cavitation -- of its inception, dynamics, and interaction with the structure and the boundary layer of the hydraulic system --
still fail to provide adequate design guidance. Full scale testing, on the
other hand, is often hampered by the large size of the structure to be
tested. Hence, model testing has been successfully employed for many
decades.

During and after WW II, cavitation has become important in the design of
anti-submarine missiles, which are exposed to serious cavitation when
entering the water at high speed. A large cavitation is usually observed to
extend from the nose to beyond the tail so that the tail fins fail to
exert stabilizing forces except through collisions with the cavity wall.
As a consequence, the missile tends to travel on a curved trajectory.

The following model study was undertaken to ascertain the model rules
governing the trajectory of a missile with a cavity attached to it. The
cavity, before it eventually collapses, is known to accompany the missile
far beyond its water entry stage. The available test facility, however,
was short, restricting the missile's water penetration to about one missile
length. This restriction was not considered a disadvantage, for the laws
governing the missile's trajectory and, hence, the model rules, would not
change during the time the cavity remained attached to the missile.

129

Consequently, if agreement between model and prototype could be achieved
during the early stage of the cavity (so the experimenter argued), it would
also obtain during later stages.

Cavitation is considered the result of a complex interplay among inertial
forces of the missile (induced by the non-linear course of the missile
and also by the sudden change of velocity at water entry) and of the water,
and pressure forces of the water, the atmospheric air, and the water vapor.
The missile's weight must be considered, too; but viscous effects may be
neglected since the water flow will be highly turbulent.

Inertial forces are governed by Newton's law of inertia and the pi-number
derived from it (called Newton number)

$$F \triangleq ma \qquad\qquad F \triangleq \rho \ell^2 v^2 \qquad Ne = \frac{F}{\rho \ell^2 v^2}$$

$$m \triangleq \rho \ell^3 \qquad a \triangleq \frac{\ell}{t^2} \qquad t \triangleq \frac{\ell}{v}$$

Weight forces are scaled in accordance with the law of gravitation and the
ensuing pi-number

$$F \triangleq mg \qquad\qquad F \triangleq g \rho \ell^3 \qquad \pi_g = \frac{F}{g \rho \ell^3}$$

$$m \triangleq \rho \ell^3$$

Pressure forces do not follow a specific law, and no principal pi-number
can be derived from them; they must be scaled in accordance with the
already derived pi-numbers.

$$p \triangleq \frac{F}{\ell^2} \qquad \frac{p}{p'} = \frac{F}{F'} \frac{\ell'^2}{\ell^2} \qquad \frac{p}{p'} = \frac{\rho}{\rho'} \frac{\ell}{\ell'}$$

$$g = g'$$

$$\pi_g \qquad \frac{F}{F'} = \frac{\rho \ell^3}{\rho' \ell'^3}$$

The Newton number and π_g can be combined to the Froude number from which a model rule for the missile's entry velocity is derived.

$$\underline{Fr} = \sqrt{\frac{\pi_g}{Ne}} = \frac{v}{\sqrt{g\ell}} \quad\longrightarrow\quad \frac{v}{v'} = \sqrt{\frac{\ell}{\ell'}}$$

The model rule for pressure, $p/(\rho\ell) = p'/(\rho'\ell')$, entails the following modeling requirements. The pressure, p, represents all pressures of importance: the atmospheric pressure, the vapor pressure of the fluid, and the hydrostatic pressure. The density, ρ, represents all densities of importance: the density of the atmosphere, the fluid, and the missile. Now, the preferred model fluid and the preferred model atmospheric gas are water and air, respectively. Since the use of water for model and prototype prescribes a density scale factor of unity, the pressure model rule simplifies to $p/p' = \ell/\ell'$. As a consequence, the pressure and hence the density of the model's atmospheric gas -- if composed of air -- will be smaller than the pressure (and density) of the prototype's air so that the density scale factor cannot be maintained unity for the atmospheric gases. It followed that either the requirement of equal atmospheric density must be violated to permit the use of air for the model, or a heavy gas must be employed, a gas that at very low pressure would have the density of atmospheric air. Such a gas was found to be Freon 114B2, with the density of atmospheric air at a pressure of only 1/9 atm. Thus, for a length scale factor of 9, the requirement of unity density scale factor for fluid and atmosphere appeared to be satisfied.

When brought in contact with water, however, the Freon gas was found to absorb a large amount of the much lighter water vapor so that, at equilibrium, the density of the Freon-water vapor mixture turned out to be 31% lower than required. To reduce water evaporation, the model water was chilled to $10°C$, which increased the density of the model's atmosphere by 10%. The remaining discrepancy of 21% was considered tolerable.

The entry cavity is filled not only with atmospheric gas but also with water vapor. Consequently, the pressure of the model's water vapor, too, should be reduced by a factor of 9 (i.e., by the length scale factor). Chilling the model's water helped bring the vapor pressure down somewhat.

But even if the temperature of the model water would have been lowered to, say, 5°C, with the prototype water at 20°C, the vapor pressure ratio would have been no larger than approximately 3, which is much smaller than the desired ratio of 9.

This obstacle was circumvented by relaxing the requirement of scaling all pressures. The inception of cavitation is usually considered to depend on the difference between the local static pressure (atmospheric pressure plus hydrostatic pressure) and the vapor pressure. If this difference is scaled, rather than the vapor pressure, then the original pressure model rules assumes the form[1]

$$\frac{(p_a + \rho_f\, gh) - p_v}{(p_a' + \rho_f'\, gh') - p_v'} = \frac{\rho l}{\rho' l'}$$

where p_a is the atmospheric pressure, ρ_f is the fluid density, h is the fluid height, p_v is the vapor pressure, ρ/ρ' is the density scale factor. For practical reasons, the density scale factor should be unity, i.e., $\rho/\rho' = \rho_f/\rho_f' = 1$, as outlined earlier. Then, with $h/h' = l/l'$, the model rule simplifies to

$$\frac{p_a - p_v}{p_a' - p_v'} = \frac{l}{l'}$$

Ideally, $p_a/p_a' = l/l'$ and $p_v/p_v' = l/l'$. If the atmospheric pressure is scaled correctly, but the vapor pressure is not, cavitation modeling will be in error, as pointed out. The error will not be large, however, because the vapor pressure is much smaller than the atmospheric pressure, at least for the prototype. For instance, the error would be only 11% with the following data:

Atmospheric pressure, p_a = 740 mm Hg

Vapor pressure at 20°C, p_v = 20 mm Hg

Length scale factor l/l' = 9

Vapor pressure of model water chilled at 10°C, p_v = 10.4 mm Hg

---

[1] A similar number is the Thoma number, σ; see Appendix B.

Then, instead of 9, the length scale factor would turn out to be

$$\frac{\rho_a - \rho_v}{\rho_a \dfrac{\ell'}{\ell} - \rho_v} = 10$$

Tests were made with full-scale aircraft dummy torpedos at the Moris Dam
Torpedo Range in Azusa, California, and with small models in the launching
tank of the Naval Ordnance Test Station in Pasadena, California. Since the
model studies were restricted to about one missile length of penetration, it
was assumed that within this regime the tail would not contact the cavity
wall and therefore the configuration of the tail would not affect the entry
pitch. Hence, for both the full-scale dummy and the models, a cylindrical
afterbody with a simple tail was used, Fig. 2.1.

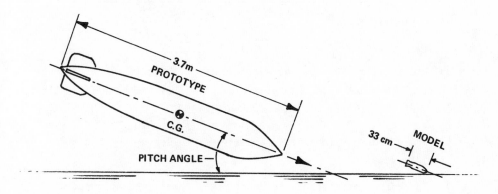

Fig. 2.1.   Missile entry into water.

To match the given model launching facility, the length scale factor was
established at 11.21 instead of at 9. The models were made of Dural; they
had internal adjustable weights to achieve properly scaled masses and moments
of inertia. The effective densities of model and protype were made the same
so that $m/m' = (\ell/\ell')^3$ and $I/I' = (\ell/\ell')^5$, where I is the moment of inertia around
the center of gravity. The water contact angle, i.e., the angle between
missile axis and horizontal plane was maintained at 20.5 deg for both model
and prototype. The model contact velocity was scaled in accordance with
the rule evolving from the Froude number. Table 2.1 lists some of the model
and prototype data.

TABLE 2.1. Model and Prototype Data.

PARAMETER	PROTOTYPE	MODEL
DIAMETER	56.95 cm	5.08 cm
WEIGHT	6788 N	4.82 N
MOMENT OF INERTIA ABOUT TRANSVERSE AXIS THROUGH C.G.	1045 kg m^2	590 x 10^2 g cm^2
WATER-CONTACT VELOCITY	122 m/s	36.4 m/s
WATER CONTACT ANGLE	20.5°	20.5°
ATMOSPHERIC PRESSURE	1 atm	0.091 atm

1 atm = 98066 N/m^2

The model atmospheric pressure was adjusted to 1/11.21 atm, in accordance with the pressure model rule. To study the effect of atmospheric density on the model test results, two tests were performed:

(1) The effect of atmospheric density was neglected, with air at 1/11 atm.

(2) The effect of atmospheric density was taken into account, with an atmosphere of Freon 114B2 and the model water chilled to 10°C.

In the second test, the launching tank was evacuated to the vapor pressure of the water and the residual air washed out with Freon, after which the tank pressure was brought to 1/11 atm (instead of the more precise pressure of 1/11.21 atm). To ensure homogeneous distribution, the gas was stirred by a small fan in the dome of the tank. Immediately before the test, the fan was stopped, and a gas sample was withdrawn to determine its density at tank pressure.

During the model's entry into the water, pitch and penetration data were optically measured at intervals of 0.4 ms to 0.8 ms. Prototype data were taken in similar fashion at intervals of 2 ms.

Model test results of the relaxed testing, test (1), and the more accurate
test (2), were compared with prototype results for three types of missile
heads -- the hemisphere, the disk-ogive, and the disk-cylinder. With the
hemisphere, disagreement between relaxed and accurate testing was negligibly
small, and both kinds of tests showed good agreement with prototype results
(see Fig. 2.2). The disk-ogive head, however, showed great sensitivity to
the relaxed testing; better agreement with prototype results was found in the
accurate testing (Fig. 2.3). The conclusion is that the inertial effects of
the atmosphere are more pronounced for the disk-ogive head.

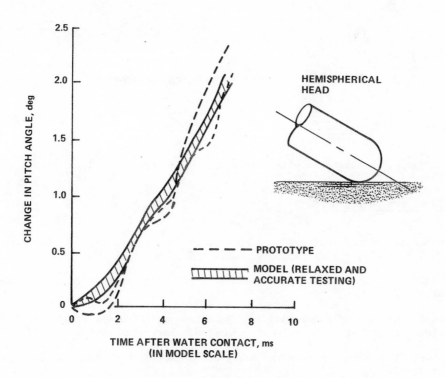

Fig. 2.2.   Pitch angle during entry — hemispherical head.

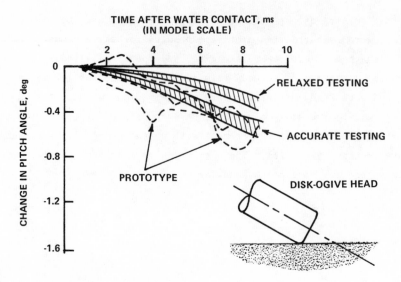

Fig. 2.3.   Pitch angle during entry — disk-ogive head.

Source of Information

J.G. Waugh, "Water-entry pitch modeling," *J. Hydronautics* 2, 2, 87-92 (Apr 1968).

Related References

Symposium on Cavitation Research Facilities and Techniques, J.W. Holl and G.M. Wood (eds), Fluids Eng. Div. Conf., ASME, New York (May 1964).

A. Thiruvengadam, "Cavitation erosion," *Appl. Mech. Reviews* 24, 3, 245-253 (Mar 1971).

Cavitation State of Knowledge, J.M. Robertson and G.F. Wislicenus (eds), Fluids Eng. and Appl. Mech. Conf., ASME, New York (Jun 1969).

A. May, "Review of water-entry theory and data," *J. Hydronautics* 4, 4, 140-142 (Oct 1970).

H.I. Abelson, "Pressure measurements in the water-entry cavity," *J. Fluid Mech.* 44, Pt. 1, 129-144 (Oct 1970).

A.J. Acosta and B.R. Parkin, "Cavitation inception -- a selective review," *J. Ship Research* 19, 4, 193-205 (Dec 1975).

See also references in Source of Information, and in the papers by A. Thiruvengadam and by J. Acosta. Numerous references can be found in the *Cavitation Forum* Conference Proceedings published annually by the ASME, Fluids Eng. Div., New York, N.Y. (10th Forum -- 1975).

CASE STUDY 3

Sedimentation

The numerous and often quite diverse phenomena caused by the interaction of
a moving fluid with a very large number of small mobile particles or, for
short, by sedimentation have long been of interest to a variety of profes-
sional people such as physicists, geologists, civil engineers, ecologists,
hydraulicists, and farmers. Many of the sedimentation phenomena are caused
by natural wind and water flow, and great efforts have been made to lessen
or prevent their adverse effects such as erosion of agricultural areas,
depositing of silt in rivers and harbors, wandering of sand dunes, and
scouring of hydraulic structures. Others are exploited on a smaller scale
in industrial processes where discrete particles must be transported, mixed,
or separated with the help of moving fluids or gases.

The mechanism of most sedimentation phenomena is not well understood. Hence,
empirical solutions prevail, based often on the clever application of scale
modeling. The following case study gives an example.

3.1 Sediment Waves

One of the factors that unfavorably influence the transport rate of fluids in
channels with sediment beds is the formation of ripples or waves (miniature
"dunes") on the bed. It seems that small irregularities in the bed give rise
to vortexes, whose increased velocity causes particles of the bed to move
out of the range of the vortex and to settle down at some distance, thus
serving to generate a new vortex, and so on. An exact analytical solution
to ripple formation is unavailable. Therefore, an attempt was made to
identify the dominant laws by scale modeling.

Stated in most general terms, the movement of loose bed material is governed by the inertial forces of the particles and of the water against them, by the weight of the floating particles, and by the viscous forces acting between water and particles. Hence, three physical laws and three pi-numbers derived from them prevail.

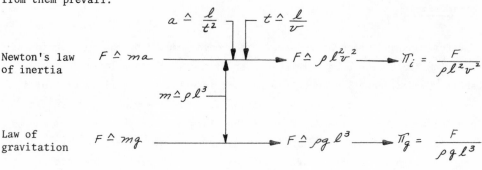

By elimination of the representative force, F, two well-known pi-numbers evolve.

Reynolds number $$Re = \frac{\pi_v}{\pi_i} = \frac{\rho v l}{\mu} = \frac{v l}{\nu}$$

Froude number $$Fr^2 = \frac{\pi_g}{\pi_i} = \frac{v^2}{g l}$$

The simultaneous identity of Reynolds and Froude numbers raises a well known problem: the length scale factor becomes a function of the scale factor of kinematic viscosity. In representative terms,

$$\left(\frac{Re}{Fr}\right)^2 = \frac{g l^3}{\nu^2} \xrightarrow{\; g = g' \;} \frac{l}{l'} = \left(\frac{\nu}{\nu'}\right)^{2/3}$$

No readily available fluids possess the kinematic viscosities that would
make a useful model-prototype relation. The author of the Source of Information considered, for instance, a 49.5% solution of glycerol for the prototype,
and water for the model. Density and viscosity were,
for the glycerol solution,

$$\rho = 1.125 \text{ g/cm}^3; \; \mu = 5.89 \text{ cp,}$$

and for water (20°C),

$$\rho' = 0.998 \text{ g/cm}^3; \; \mu' = 1.00 \text{ cp.}$$

Therefore,

$$\frac{\ell}{\ell'} = \left(\frac{\nu}{\nu'}\right)^{2/3} = \left(\frac{\mu \rho'}{\rho \mu'}\right)^{2/3} \approx 3, \text{ a rather small value.}$$

The density scale factor of $\rho/\rho' = 1.127$ applies not only to the fluid but
also to the sediment. The author considered coal particles as prototype
sediment ($\rho = 1.26$ g/cm^3) and Araldite epoxy resin ($\rho' = 0.999$ g/cm^3) as
model sediment. The necessary speed ratio follows from the Reynolds number
as $v/v' = (\nu/\nu')(\ell'/\ell) = \sqrt{\ell/\ell'} = 1.73$. With these results, model tests could have
been initiated. Instead, the author decided to investigate the possibility
of performing more realistic test work with water, not glycerol, as prototype
fluid.

The use of the same fluid for model and prototype calls for less than perfect
similarity, for with $\nu = \nu'$, Reynolds and Froude numbers would result in conflicting model requirements. This impasse could be avoided, however, if the
inertial forces of the sediment were much smaller than the rest of the forces
and could therefore be neglected. Then, Newton's law of inertia had only to
be applied to the fluid, and π_i would assume the form of

$$\pi_{iw} = \frac{F}{\rho_w \, \ell^2 v^2}$$

A further simplification, without loss of generality, can be achieved by
restricting the law of gravitation to the weight difference of water and sediment,

$$F \triangleq (\rho_s - \rho_w) \, g \, \ell^3 \longrightarrow \pi_{fs} = \frac{F}{(\rho_s - \rho_w) \, g \, \ell^3}$$

With these two modifications, a modified Reynolds number and a modified Froude number evolve (the latter is often called densimetric Froude number).

$$\pi_1 = \frac{\pi_v}{\pi_{i w}} = \frac{\rho_w v l}{\mu} \quad \text{and} \quad \underline{Fr}_d = \frac{\pi_{fs}}{\pi_{iw}} = \frac{\rho_w v^2}{(\rho_s - \rho_w) g l}$$

From them, two model rules are developed, leading to a relation between the length scale factor and the sediment densities.

Hence, the length scale factor is freed from its dependence on kinematic viscosity, a very desirable result since it permits the use of water for model and for prototype. Of course, a price had to be paid -- the motions of the particles had to be restricted to a state of rolling and sliding, with small or no inertial forces acting on them. The model would be invalid for higher speeds where the particles would begin to leave the bed and be carried upward ("saltation"). This restriction is no disadvantage here, however; sediment waves begin to form almost simultaneously with the initiation of particle movement, even if the movement is slow.

The model tests were performed in a tilting flume 18 m long and 65 cm wide. The model sediment was sand with a density of $\rho'_s = 2.65$ g/cm^3; the prototype sediment, Polystyrene with a density of $\rho_s = 1.035$ g/cm^3; the fluid, water with a density of 1.00 g/cm^3. With these data, $l/l' = \sqrt[3]{1.65/0.035} = 3.6$. The average grain diameter of the sand was 0.375 mm. Therefore, the grain diameter of the Polystyrene sediment was made 1.35 mm.

The same flume was used for model and prototype tests, since it was assumed that the formation of sediment waves is a two-dimensional process.

Besides the sediment density and sediment diameter, the speed of the flowing
water had to be scaled in accordance with the earlier developed modified
Reynolds number π_1 , so that $v/v' = l'/l = 1/3.6$. The necessary speeds were
achieved by tilting the flume by a very small angle. Presumably, a very
small angle would not affect the formation of waves, but it would induce
the water to flow because of gravity. Model and prototype tilting angles were
computed from the representative gravity component in flow direction,
$F \triangleq \rho_{wr} g l^3 \alpha$, where α is the tilt angle. Hence, a new pi-number was introduced:
$\pi_t = F/(\rho_{wr} g l^3 \alpha)$. Combined with π_{iw} , it resulted in the requirement of
$\pi_2 = \pi_{iw}/\pi_t = g l \alpha/v^2$ or, with $v/v' = l'/l$, in $\alpha/\alpha' = (l'/l)^3$.

For the prototype flume, a tilt angle of α = 1/5000 was elected. Consequently, α' = 3.6^3/5000 = 1/107. This small angle was assumed to have no
effect on sediment wave formation.

Finally, the flow height and the rate of flow volume had to be scaled. For
two-dimensional flow, $\dot{Q} \triangleq v b h$, where b is the flume width, and h is the flow
height. For constant flume width,

$$\frac{\dot{Q}}{\dot{Q}'} = \frac{v h}{v' h'} \longrightarrow \dot{Q} = \dot{Q}'$$

$$\pi_1 \longrightarrow \frac{v}{v'} = \frac{l'}{l} \qquad \frac{h}{h'} = \frac{l}{l'}$$

All model and prototype design parameters are summarized in Table 3.1.

TABLE 3.1. Model and Prototype Design Parameters for Approximate Similarity.

	MODEL	PROTOTYPE
FLUID	WATER	WATER
WATER FLOW RATE, cm^3/s	$\dot{Q}$	$\dot{Q}$
WATER FLOW HEIGHT, cm	h	3.6 h
FLUME TILT ANGLE, tan α	1/107	1/5000
SEDIMENT	SAND	POLYSTYRENE
SEDIMENT DIAMETER, mm	0.375	1.350
SEDIMENT DENSITY, g/cm^3	2.650	1.035

Under the influence of the flowing water, the bed material formed into ripples or waves; their spacing and motion were measured (Fig. 3.1). At very
low water speed, no wave formation occurred. At a certain, still low, speed,
a few widely spaced sediment ripples appeared, to concentrate at higher
speeds, and to disperse again at still higher water speeds.

A correctly designed model should confirm measured prototype data in scaled
fashion. Here, the wave spacing and the flow velocity were measured and
plotted, the wave spacing in nondimensional form as λ/d (λ is wave spacing,
d is grain diameter), the flow velocity also in nondimensional form as
v/v_{crit}, where v_{crit} is the threshold speed at the onset of particle movement.
The good agreement between model and prototype data in Fig. 3.2 confirms the
correctness of the assumptions on which the model design was based.

Source of Information

Ministry of Technology, Hydraulics Res. Station, *Similarity in sediment
transport by currents* by M.S. Yalin, Hydr. Res. Paper No. 6, Wallingford,
Berks., England (Feb 1965).

Related References

W.A. Price and M.P. Kendrick, "Field and model investigation into the reasons
for siltation in the Mersey estuary," *Inst. Civil Engrs. Proc.* 24, 473-518
(Apr 1963).

J.L.Bogardi, "Incipient sediment motion in terms of the critical mean
velocity," *Acta Technica Academiae Scientarium Hungaricae* 62, 1-2, 1-24
(1968).

A.M.Z. Alam and J.F. Kennedy, "Friction factors for flow in sand-bed channels,"
Proc. ASCE,J. Hydr. Div. 95, HY 6, 1973-1992 (Nov 1969).

J. Bogardi, "Hydraulic similarity in sediment transport," Ch. 2.5, *Sediment
Transport in Alluvial Streams*, Akademiai Kiado, Budapest, 1974.

P. Novak and C. Nalluri, "Sediment transport of smooth fixed bed channels,"
Proc. ASCE, J. Hydr. Div. 101, HY 9, 1139-1154 (Sep 1975).

Publications on modeling sediment transport are reviewed monthly in *Civil
Eng. Hydraulic Abstr.* (formerly *Channel*).

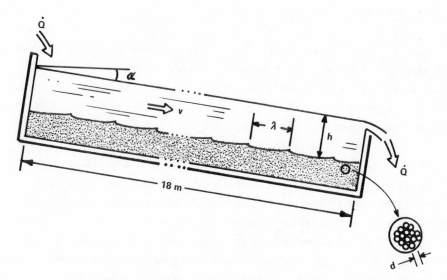

Fig. 3.1.   Tilting flume with sediment waves.

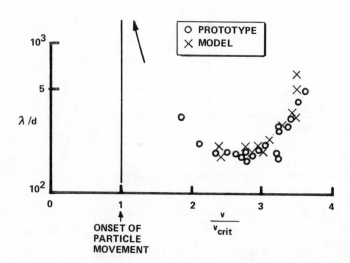

Fig. 3.2.   Dimensionless sediment wave length vs. dimensionless flow velocity; test results.

3.2 Sedimentation in Tanks by Continuous Flow of Fluid

Many kinds of industrial and municipal plants for treating water or sewage
use continuous-flow sedimentation tanks to separate solids from liquids.
But the mechanics of the separation process are not well understood. Hence,
scale model experiments of the process were attempted for the purpose of
improving tank design.

The movements of particles suspended in flowing fluid are governed by the
inertial forces of particles and fluid, by the viscous forces between particles and fluid, and by the buoyant forces of the particles. Therefore, the
Reynolds number and the Froude number apply, as explained in the previous
case study. The Froude number can again be restricted to the weight difference between water and sediment. Instead of expressing the two pi-numbers
in terms of velocity, which is cumbersome to measure, they can be expressed
in terms of the water discharge rate, $\dot{Q}$. If, in addition, the same fluid --
water -- and the same sediment material are used for model and prototype,
then

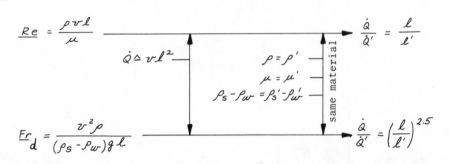

$$Re = \frac{\rho v \ell}{\mu}$$

$$\dot{Q} \triangleq v \ell^2$$

$$\rho = \rho'$$
$$\mu = \mu'$$
$$\rho_s - \rho_w = \rho_s' - \rho_w'$$

same material

$$\frac{\dot{Q}}{\dot{Q}'} = \frac{\ell}{\ell'}$$

$$Fr_d = \frac{v^2 \rho}{(\rho_s - \rho_w)g\ell}$$

$$\frac{\dot{Q}}{\dot{Q}'} = \left(\frac{\ell}{\ell'}\right)^{2.5}$$

The two rules derived from Reynolds and Froude spell out conflicting model
design requirements; which to neglect can be decided only by experiments. If
the discharge rate $\dot{Q}$ of both tanks (one to be the prototype and the other
the model) is measured, and if $\dot{Q}/\ell$ for the two tanks is plotted versus a
dimensionless quantity of interest (the ratio c_e/c_i, for instance, where c_e
and c_i are concentrations of sediments in the effluent and the influent),
and if all data collapse into one curve, then we recognize that the inertial
and viscous forces dominate and the buoyant force can be neglected. On the
other hand, if plots of $\dot{Q}/\ell^{2.5}$ versus c_e/c_i collapse, then we know that
inertial and buoyant forces dominate and viscous forces can be neglected.

In the third possibility, that of viscous and buoyant forces being dominant over inertial forces, we form a pi-number from $\underline{Re}$ and $\underline{Fr}_d$

$$\pi_3 = \frac{Fr_d}{Re} = \frac{\mu v}{(\rho_s - \rho_w) g \ell^2} \quad\xrightarrow{\hspace{3cm}}\quad \frac{\dot{Q}}{\dot{Q}'} = \left(\frac{\ell}{\ell'}\right)^4$$

$$\dot{Q} \triangleq v\ell^2 \qquad \begin{array}{l} \text{same} \\ \text{material} \end{array} \begin{cases} \mu = \mu' \\ \rho_s = \rho_s' \\ \rho_w = \rho_w' \end{cases}$$

In this case we expect that plots of $\dot{Q}/\ell^4$ versus c_e/c_i for model and prototype collapse.

Two geometrically similar rectangular sedimentation tanks were constructed (length scale factor of 3) with horizontal bed and simple inlet and outlet weirs (inset, Fig. 3.3). Powder mixed into the influent water provided sedimentation. With various combinations of sediment concentrations and flow rates, both the influent and effluent sediment concentrations were measured.

The results of tests with the two tanks are plotted in three ways: $\dot{Q}/\ell$, $\dot{Q}/\ell^{2.5}$, and $\dot{Q}/\ell^4$, all versus c_e/c_i, where the characteristic length, ℓ, is taken to be unity for the model; and three, for the prototype. As Fig. 3.3 indicates, none of the three prototype curves collapses with the model curve. The disagreement could be caused by experimental errors or by incorrect scaling, but a good possibility is that all three forces -- viscous, inertial, and gravitational -- participate equally in the phenomenon.

A closer examination of the Source of Information revealed that the model was indeed scaled incorrectly: the powders used for model and prototype were identical not only in material but also in grain size. For a scale model experiment to be valid, however, the particles, too, must be geometrically scaled. Hence, this case study demonstrates clearly that the purpose of model testing may be defeated unless careful attention is given to all important details.

Source of Information

D.M. Thompson, "Scaling laws for continuous flow sedimentation in rectangular tanks," *Proc. Instn. Civil Engrs*. 43, 453-461 (Jul 1969); Discussion: *Proc. Instn. Civil Engrs*. 46, 387-393 (Jul 1970).

Related References

R.J. TeKippe and J.L. Cleasby, "Model studies of a peripheral feed settling tank," *Proc. ASCE, J. Sanitary Eng. Div*. 94, SA1, 85-102 (Feb 1968).

M.S. Clements and A.F.M. Khattab, "Research into time ratio in radial flow sedimentation tanks," *Proc. Instn. Civil Engrs*. 40, 471-494 (Aug 1968).

B.W. Gould, "Sedimentation models for hydraulic design of clarifiers," *Inst. Engrs. Australia, Civil Eng. Trans*. CE 11, 1, 55-59 (Apr 1969).

G.A. Price and M.S. Clements, "Some lessons from model and full-scale tests in rectangular sedimentation tanks," *J. Instn. Water Pollut. Control* 73, Pt. 1, 102-113 (1974).

3.3 Localized Scour

Scour in stream beds occurs when water, moving swiftly past an obstruction such as a bridge pier, generates strong vortices that scoop cavities in the stream bed. Steady-state scour is the term used to describe the continuous scooping of bed material and its continuous replacement with sediment from the edges and upstream slope of the cavity.

Perfect scaling is frustrated by the severe restrictions imposed by the simultaneous identity of Froude and Reynolds numbers, i.e., by the equal importance assigned to (1) the inertial forces of the particles, (2) the inertial forces of the water, (3) the buoyancy forces of the floating parti- cles, and (4) the viscous forces between fluid and particles. Therefore, to simplify and thus make scale modeling possible, the experimenter of the Source of Information assumed the velocity of the water involved in scouring

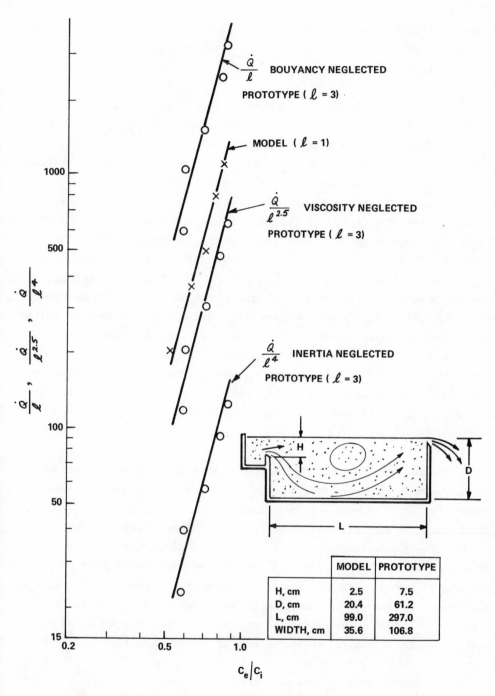

Fig. 3.3.   Normalized flow rate (recalculated from original data) vs. ratio of sediment concentrations.

$\dot{Q}$ in cm³/s;   x = model (ℓ = 1);   o = prototype (ℓ = 3)

to be high, high enough to induce turbulence and, hence, high inertial forces
so that point (4) could be neglected. Consequently, only (1) through (3)
needed to be considered.

As derived in the first example ("Sediment waves"), the mechanisms of inertia
and buoyance lead to the densimetric Froude number $\underline{Fr}_d = \frac{v^2 \rho}{(\rho_s - \rho_f) g \ell}$, where
ρ_f is the density of the fluid, and ρ represents both the density of the
sediment and of the fluid. Hence, to keep the density scale factor constant,
both the density of the sediment and the density of the fluid had to be
changed simultaneously. The density of the sediment can be changed easily by
a large factor; the density of the fluid cannot. The author decided therefore
to try to use water for both model and prototype, thereby neglecting the
inertial effects of the sediment, i.e., point (1). Then, the densimetric
Froude number simplifies to $\underline{Fr}_d = \frac{v^2}{(\rho_s/\rho_w - 1) g \ell}$. Figure 3.4 shows the simple
test setup schematically. A pipe of transparent plastic, 7.5 cm in diameter,
was fitted with a transparent, sloped cavity and a piston to feed it continuously with sediment. Water was the working fluid in all tests, whereas
the type and size of the sediment material were varied (Table 3.2). During
a run, the water would form a vortex within the scour hole which, at a certain minimum flow speed (and beyond), would pick up sediment and transport
it downstream. In all runs, the rate of water discharge and the rate of
sediment removal and replacement were adjusted to steady-state conditions.

About 150 runs were made and the results plotted in terms of the nondimensional sediment transport rate, $\dot{Q}_s/(v D_s^2)$, versus the densimetric Froude
number, $\underline{Fr}_d$ (both numbers were recalculated from some of the data given in the
first Source of Information), where $\dot{Q}_s$ is the sediment transport rate, v
is the water flow velocity in the main pipe, D_s is the diameter of the sediment grains. In the plot, Fig. 3.5, the data scatter considerably, a result,
perhaps, of the violation of one of the fundamental requirements of scale
modeling -- to maintain the same length scale factor for all important
dimensions. In the experiments, the size of the pipe and the size of the
cavity remained unchanged, while the size of the sediment grains changed
greatly. In the author's opinion, no firm conclusions about the validity of
the proposed model rules can be made without the same length scale factor for
both sediment grains and scour cavity (apparently unimportant is the main
pipe diameter which thus might be kept constant).

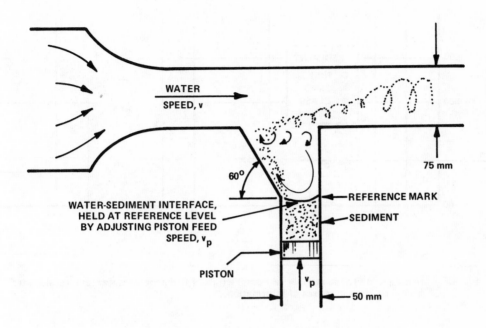

Fig. 3.4. Schematic of scour model.

TABLE 3.2. Some of the Sediments Tested.

SEDIMENT	MATERIAL	DENSITY g/cm³	GRAIN DIAM. D_s, mm
1	NICKEL	8.75	0.570
2	SAND	2.63	0.185
3	GLASS	2.46	0.106
4	LUCITE	1.20	0.250

Source of Information

A.R. LeFeuvre, "Sediment transport functions with particular emphasis on localized scour," Ph.D. Diss., Civil Eng., Dept., Georgia Inst. of Tech., Atlanta, 1965.

Dissertation discussed in:

M.R. Carstens, "Similarity laws for localized scour," *Proc. ASCE, J. Hydr. Div.* 92, HY3, 13-36 (May 1966). Discussions: *Proc. ASCE, J. Hydr. Div.* 93, HY2, 67-71 (Mar 1967) and 94, HY1, 303-306 (Jan 1968).

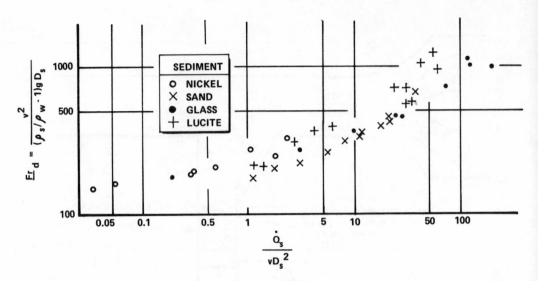

Fig. 3.5.   Densimetric Froude number vs. nondimensional sediment transport rate (recalculated from data given in source of information).

Related References

H.W. Shen et al, "Local scour around bridge piers," *Proc. ASCE, J. Hyd. Div.* 95, HY6, 1919-1940 (Nov 1969).

Univ. Iowa, Inst. Hydr. Res., *Scale Effects in Hydraulic Model Tests of Rock Protecting Structures* by E.M. O'Loughlin et al, IIHR 124, Iowa City, Iowa, Feb 1970.

South Dakota Dept. of Highways, *Experimental Study of Reducing Scour Around Bridge Piers Using Piles* by F.M. Chang and M. Karim, Jan 1972.

A.K. Tyagi, "Modeling of local scour around spur dykes in streams," *Proc. Intl. Assoc. for Hydr. Res., Intl. Symp. on River Mechanics* 1, pp. 115-165, 1973.

H. Kikkawa and S. Fukuaka, "Study of localized scour around a bridge pier and its prevention," ibid. pp. 105-116.

G. Glazik, "Hydraulic scale model tests on local scour near offshore structures under wave action," *Proc. 16th Intl. Assoc. for Hydr. Res., Congr. on Fundamental Tools to be Used in Environmental Problems* 2, pp. 189-195, 1975.

C.J. Posey, "Tests of scour protection for bridge piers," *Proc. ASCE, J. Hydr. Div.* 100 HY12, 1773-1783 (Dec 1974). Discussions: *Proc. ASCE, J. Hydr. Div.* 101, HY10, 1369-1371 (Oct 1975), and HY11, 1454-1455 (Nov 1975).

Publications on modeling scour are reviewed monthly in *Civil Eng. Hydr. Abstr.* (formerly *Channel*).

CASE STUDY 4

Urban Air Pollution

Man lives on the bottom of an atmospheric ocean. His life is greatly affected
by atmospheric motions near the ground, particularly in an industrial city
where wind forces on tall bulidings, dispersion of air pollutants, and heat
released by factories and other man-made energy sources make life less plea-
sant than in a rural area.

The atmospheric boundary layer (the surface layer up to a few hundred meters
altitude) with its complex and rapidly changing interactions of buoyancy
forces, drag forces, inertial forces, topographic features, buildings, parks,
surface heat release...is almost impossible to reproduce mathematically with
reasonable accuracy. As a result, wind tunnel studies supported by field
studies have recently become a major source of information.

Scaled simulation of the atmospheric boundary layer in a wind tunnel cannot,
of course, deliver exact results (no simulation can); but despite some unavoid-
able model simplifications, good agreement between wind tunnel and field data
has been achieved, as demonstrated in the following case study.

Fort Wayne, a typical, large industrial city in Indiana, USA, proved to be
ideally suited for a combined wind tunnel and field study because extensive
prototype data were available, and because the city is located in a flat rural
environment, a feature easy to model in a wind tunnel. In the prototype
experiments, an airplane flying upwind across the wind at an altitude of
about one hundred meters had continuously released fluorescent pigments, and
the dosages reaching the surface had been measured by many stations distributed over the city. This experiment was to be repeated in an atmospheric
wind tunnel,[1] to confirm or modify the applied model technique. It was

---

[1] An atmospheric wind tunnel is different from a regular wind tunnel in that
it is operated at low speed with a large boundary layer at the bottom.

151

assumed at the outset that the average horizontal wind speed would be constant so that only vertical changes had to be modeled. Also, earlier work indicated that for areas with a characteristic distance of less than 150 km (for Fort Wayne, it was 13 km), effects of the earth's rotation upon the air's inertia may be disregarded.

A city feature whose influence on the dispersion pattern of pollutants was uncertain and needed be explored was the so-called heat-island effect. A city, particularly an industrial city with its many artificial heat sources, releases large amounts of heat into the atmosphere, which may significantly affect the distribution of air pollutants.

Atmospheric motions above a heated area are governed by forces of buoyancy, inertia, and viscosity. Hence, three laws apply.

Newton's law of inertia

$$F \triangleq \rho \ell^3 \frac{\ell}{t^2} \qquad \pi_i = \frac{F}{\rho \ell^2 v^2}$$

$$t \triangleq \frac{\ell}{v}$$

Newton's law of gravitation, expressed as buoyant force

$$F \triangleq \Delta \rho \ell^3 g \qquad \pi_f = \frac{F}{\Delta \rho \ell^3 g}$$

$$\tau \triangleq \frac{F}{\ell^2}$$

Shear stress of Newtonian fluid

$$\tau \triangleq \mu \frac{v}{\ell} \qquad \pi_V = \frac{F}{\mu v \ell}$$

where $\Delta \rho$ is the representative density difference between heated and unheated air.

From the three laws, two well-known pi-numbers are derived

Reynolds number $\underline{Re} = \dfrac{\pi_V}{\pi_i} = \dfrac{\ell v}{\nu}$, with $\nu \equiv \dfrac{\mu}{\rho}$ (kinematic viscosity)

Bulk Richardson number

$$\underline{Ri} = \frac{\pi_i}{\pi_f} = \frac{\Delta \rho g \ell}{\rho v^2}$$

Exact similarity would demand that $\Delta \rho \triangleq \rho$ so that the bulk Richardson number would change into the Froude number, $\underline{Fr} = \frac{v}{\sqrt{g\ell}}$. Retaining $\Delta \rho$ and ρ, however, led to useful simplifications.

Variations of air density above a city are caused predominantly by temperature changes; the influence of pressure can be excluded because air pressure above a city is usually constant. Then, with the equation of state for ideal gas,

$$\underline{Ri} = \frac{\Delta \rho g \ell}{\rho v^2} \longrightarrow \underline{Ri} = \frac{\Delta \theta g \ell}{\theta v^2} \longrightarrow \frac{\Delta \theta}{\Delta \theta'} = \frac{\ell'}{\ell}\left(\frac{v}{v'}\right)^2$$

Equ. of state $\quad \frac{\Delta \rho}{\rho} = \frac{\Delta \theta}{\theta} \quad\rfloor \qquad\qquad \theta = \theta' \quad\rfloor$
(p = const)

$\Delta \theta$ represents the heat-island differential, and θ, the average absolute temperature.

The last scale-factor relation is derived under the assumption that average absolute prototype temperature equals the average absolute model temperature. This is nearly true even if the model's heat-island differential, $\Delta \theta'$, is higher than that of the prototype's, $\Delta \theta$. A higher $\Delta \theta$ cannot be avoided, however, for the following reason.

To accommodate the model of a large city in a wind tunnel, the length scale factor must be selected in the order of 2000 or more. Since, according to the Reynolds number, $v/v' = \ell'/\ell$ (with $\nu = \nu'$), an impossibly large wind tunnel speed would be required. Fortunately, an atmospheric boundary layer is always turbulent; inertial forces are always much larger than viscous forces; therefore, viscous forces and with them the Reynolds number need not be considered. Consequently, the wind velocity of the model can be selected at a low level that would require only moderate power for the wind tunnel blower. In addition, a low wind velocity would bring the scale factor of the heat-island temperature differential, $\frac{\Delta\theta}{\Delta\theta'}$, closer to unity and, hence, closer to perfect temperature similarity (which is defined as $\frac{\Delta\theta}{\Delta\theta'} = \frac{\theta}{\theta'}$). On the other hand, the model velocity should be high enough to ensure "rough" flow. Model tests of pipe flow have shown that at low Reynolds number (that is at low speed), the flow may not be fully turbulent; viscous forces may still play a significant role. Viscous forces can be suppressed, however, by making the walls physically rough, for instance by coating them with sand particles. Figure 21 in Part I shows that the minimum speed required to initiate "rough" flow depends on the size of the roughness elements: the larger the elements, the lower the speed at which the flow becomes independent of viscous forces. Similar considerations were applied to the city model. The city buildings were considered roughness elements, and the question was investigated of how high they would have to be made (in the average) to stimulate rough flow.

The minimum speed the wind tunnel was able to provide was 2.44 m/s. It was found, by using various empirical equations of rough flow over plates developed elsewhere, that at this speed, an average building height of 3.8 mm was necessary to establish rough flow.[1] Since the average height of the prototype buildings was about 7.6 m, a length scale factor of 7600/3.8 = 2000 was required. Unfortunately, the wind tunnel was not large enough to accommodate a model of the required size. Fort Wayne covers an area of approximately 13 x 13 km, and the usable width of the tunnel was 3.65 m. Consequently, the length scale factor was fixed at about 4000. In view of this dilemma, it was decided to use two length scale factors, a horizontal one of $\ell/\ell' = 4000$, and a vertical one of $\jmath/\jmath' = 2000$. Whether or not the use of two length scale factors would seriously impair the predictive power of the model could only be verified experimentally by comparing model and prototype results.

To compute the heat-island differential of the model, $\Delta\theta'$, from the bulk Richardson number, the upwind speed, ν, and the heat-island differential of the prototype, $\Delta\theta$, had to be determined. From a 30 m tower in downtown, Fort Wayne, $\Delta\theta$ was found to be about 0.28°C. The upwind speed was measured at approximately $\nu = 7.9$ m/s. With these data,

---

[1]The applied technique is analog to the trip-wire technique, see Section "Transition fixing," Part I. In both cases, viscous forces are suppressed in favor of inertial forces.

$$\Delta\theta' = \Delta\theta \frac{z}{z'}\left(\frac{v'}{v}\right)^2 = 0.28 \cdot 2000 \cdot \left(\frac{2.44}{7.9}\right)^2 = 53.4\,°C$$

Experiments. Four major parts of the prototype had to be modeled in accordance with the previously developed criteria: the city and its environment, the heat island, the approach flow, and the release of pollutants.

City blocks were cut out of Masonite sheets; the surrounding rural areas, of coarse sandpaper.

The heat-island effect was modeled by placing Nichrome wires over the model's surface and applying a voltage that would generate a temperature differential of 53.4°C.

The release of pollutants from an airplane was simulated by a tubing traveling across the wind tunnel and releasing tracer gas Kr-85[1] in the direction of flow. The height of the release could be adjusted to the scaled height of the aircraft (appr. 75 m). Samples were drawn at about 1 mm above ground at 25 stations distributed over the city model and collected in separate glass bottles. After a run, the dosage, D, of each bottle was measured by a Geiger-Müller counter in $\mu ci\cdot min/cm^3$,[2] and related to a release dosage, D_0, computed from the concentration of the release gas, c, in $\mu ci/cm^3$; the gas volume released per second, $\dot{V}$, in cm^3/sec; the time needed for a trip of the tubing across the tunnel, t, in min; the wind velocity at the release, v, in cm/sec; the height of the release, h, in cm; and the length of the release line, d, in cm. With these quantities, the dimensionless dosage was defined as

$$\mathbb{D} = \frac{D}{D_0} \quad ; \quad D_0 = \frac{c\dot{V}t}{dvh} \quad , \quad in \quad \frac{\mu ci\cdot min}{cm^3}$$

----

[1] A beta-emitting radioactive gas.

[2] The curie (abbreviation, ci) is the unit of radioactivity; 1 ci = 3.70×10^{10} disintegrations per second.

For prototype runs, a dimensionless dosage was computed by relating the
particle dosage of the sample bottles to the released particle dosage. In
this way, prototype and model dosages could be compared.

The approach flow is characterized by its vertical velocity and its turbulence
distribution; longitudinal and lateral distributions are assumed to be constant. The velocity profile of the prototype was approximately known from
tower and balloon data. This profile was then reproduced in the wind tunnel
by placing a longitudinal grid of cardboard tubes at the entrance section in
two layers, Fig. 4.1. Figure 4.2 indicates the achieved agreement. Turbulence intensity was measured (with hot-wire sensors) in terms of $\sqrt{\bar{u}^2}/U$,
where $\sqrt{\bar{u}^2}$ is the root-mean-square value of the fluctuating component, and U
the average velocity at a given point. With only two prototype data points
at lower heights, agreement at higher heights could not be ascertained,
Fig. 4.3.

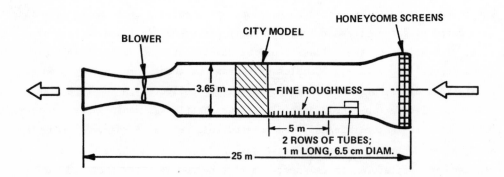

Fig. 4.1.   Environmental wind tunnel at Colorado State University, with stable air speeds
between 2.44 - 18.3 m/s. See also Fig. 6, Part I.

Results. Of the results presented in the Source of Information, we quote the
longitudinal distribution of dimensionless surface dosages measured over the
model city and prototype. Figure 4.4 shows that at about 6 km from the source,
field data and wind tunnel data do agree fairly well, despite the many approximations and simplifications introduced into the model. Wind tunnel tests
were run with and without heat island. The slope of the dosage curve with
heat island is almost the same as the slope of the field data although the
model dosages are somewhat larger. Since the slope indicates the decline of

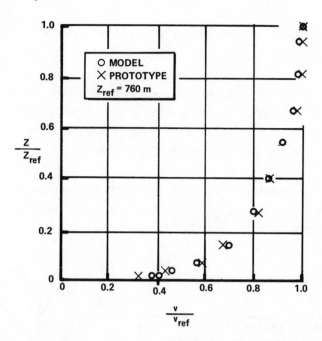

Fig. 4.2.   Vertical velocity profiles of model and prototype, measured upwind.

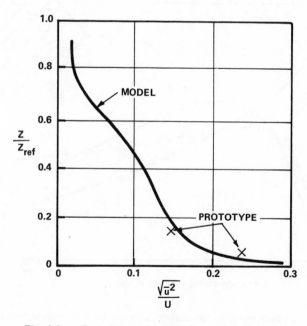

Fig. 4.3.   Turbulence intensity of model and prototype.

pollution dosages due to effects caused by the city, one must conclude that the island has indeed a marked effect on the pollution pattern.

In the regime close to the release line, agreement does not exist. It was conjectured that the downdraft of the airplane (which could not be reproduced in the wind tunnel) may have transported tracer material faster to the ground than the (horizontal) release mechanism of the model.

All in all, the wind tunnel results were considered encouraging, and further tests were planned to investigate the influence of tall buildings, parks, highways, residential and industrial areas.

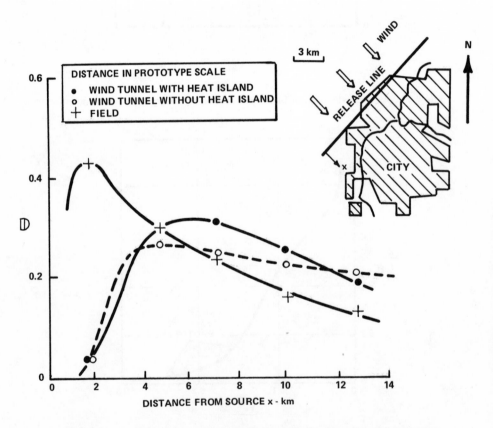

Fig. 4.4.   Dimensionless surface dosage, $\mathbb{D}$, along the direction of transport, x.

Source of Information

Colorado State Univ., Fluid Dynamics and Diffusion Lab., *Windtunnel Modeling of Flow and Diffusion over an Urban Complex* by F.H. Chaudhry and J.E. Cermak, Project THEMIS, TR-17, May 1971.

Selected Related References

J.S. Turner, "Model experiments relating to thermals with increasing buoyancy," *Quart. J. Roy. Meteorol. Soc.* 89, 379, 62-74 (Jan 1963).

G.H. Strom, "Simulating atmospheric processes in a wind tunnel," *On Atmospheric Simulation: A Colloquium*, G.M. Hidy (ed), NCAR-TN-22, Natl. Ctr. for Atmospheric Res., Boulder, Colo., 193-204 (Nov 1966).

J. Armitt and J. Counihan, "Simulation of the atmospheric boundary layer in a wind tunnel," *Atmospheric Environment* 2, 1, 49-71 (Jan 1968).

S. Nemoto, "Similarity between natural local wind in the atmosphere and model wind in a wind tunnel," *Papers in Meteorology and Geophysics*, Meteorol. Res. Inst., Tokyo, 19, 2, 131-230 (Jul 1968).

S.P.S. Arya and E.J. Plate, "Modeling of the stably stratified atmospheric boundary layer," *J. Atmos. Sci.* 26, 4, 656-665 (Jul 1969).

J.E. Cermac, "Laboratory simulation of the atmospheric boundary layer," *AIAA J.* 9, 9, 1746-1754 (Sep 1971).

W.H. Snyder, "Similarity criteria for the application of fluid models to the study of air pollution meteorology," *Boundary-Layer Meteorology* 3, 113-133 (1972).

T.R. Sundaram, G.R. Ludwig, and G.T. Skinner, "Modeling of the turbulence structure of the atmospheric boundary layer," *AIAA J.* 10, 6, 743-750 (June 1972).

Wind engineering research has expanded rapidly during the past years. A list of model tests performed since 1972 would fill many pages. Information on past and present modeling projects and programs can be found in:

Wind Engineering Digest Univ. Hawaii, Honolulu, Vol. 1 (1974) -- .

WERC (Wind Engineering Research Council) Newsletter 1, 1 (May 1975) -- .

Industrial Aerodynamic Abstracts 1, 1 (1970) -- .

CASE STUDY 5

Soil-Machine Interactions

The mechanical properties of soil have always been of importance to man --
to the farmer, the builder, the soldier, and to anybody who must till land,
dig a hole, or simply walk across a field. But as yet, no dependable engineering description of the material "soil" has come forth. Hence, when
developing earthworking equipment, we must still rely on empirical methods
and cut-and-try procedures. Under these circumstances, model tests have
proved their usefulness, particularly in the design of heavy machinery (such
as bulldozers, graders, large off-road vehicles), which is cumbersome to
handle and costly to modify when tested in full scale.

Of course, even model tests require some formal description of the working
material "soil," but the description need not go beyond very general, albeit
quantitative, statements about the fundamental mechanism of the system at
hand, as pointed out repeatedly in this book. The fundamental mechanism of
soil-machine systems is ruled by six basics: (1) inertia of soil particles;
(2) friction between soil particles; (3) cohesion between soil particles;
(4) soil weight; (5) soil elasticity; and (6) adhesion between soil and
machine. Friction, cohesion, adhesion, and elasticity all depend on the
strain, the strain rate, the geometry of the soil-machine interface, and the
degree of soil compaction during the process of soil displacement by the
interacting machine.

In view of this complexity, simplifications are necessary. Usually, when
disturbed by a piece of earthworking machinery, soil elements experience
large displacements from their original positions, mostly far beyond the
limits of soil failure. Consequently, elastic and strain effects, which
are prominent only for small displacements, can safely be neglected. Strain
rate effects, i.e., the influence of deformation speed on soil properties,
are sometimes claimed to be important; in most cases, however, they are
quite small and can be ignored. Finally, adhesion forces are usually

disregarded by assuming that a thin layer of soil adheres firmly to the con-
tacting machine surface.

With the effects of elasticity, strain, strain rate, and adhesion eliminated,
only the basics (1) through (4) survive together with the possible influence
of soil compaction. Inertia and weight of the soil are governed by Newton's
laws

$$t \triangleq \frac{\ell}{v}$$

Newton's law $\quad F \triangleq \rho \ell^3 \frac{\ell}{t^2} \longrightarrow \Pi_i = \frac{F}{\rho \ell^2 v^2}$
of inertia

Newton's law $\quad F \triangleq \rho g \ell^3 \longrightarrow \Pi_g = \frac{F}{\rho g \ell^3}$
of gravitation

Friction and cohesion of soil particles are governed by Coulomb's law if the
soil is disturbed beyond failure. Assuming the state of failure to be the
dominant feature, we have

Coulomb's law for $\quad F \triangleq c \ell^2 \longrightarrow \Pi_c = \frac{F}{c \ell^2}$
cohesive force

Coulomb's law
for internal $\quad F \triangleq F \mu \longrightarrow \Pi_\mu = \mu$
friction force

where c is the cohesion and μ, the internal coefficient of friction. The
internal coefficient of friction is a pi-number itself.

Experience shows that the soil properties of cohesion (c), internal friction (μ), and density (ρ) cannot, or can only to a very limited degree, be
selected independently; for each soil, the three quantities are given as a
package. The pi-number Π_μ, however, demands that $\mu = \mu'$. This requirement
can be satisfied if the same soil is used for the model as for the prototype,

so that $\mu = \mu'$. But then, by necessity, $c = c'$, and $\rho = \rho'$. Under these circumstances, the two pi-numbers π_g and π_c are in conflict. Hence, relaxations are required.

Natural soils can be roughly divided into two groups -- cohesive or clayey soils, and granular or sandy soils. Clay is composed of very small particles, having the size of a few microns. Consequently, the internal surface of clay is very large, and surface forces such as the cohesive force are much larger than the gravitational force. Therefore, π_g may be neglected. In addition, many clayey soils show very little internal friction so that π_μ can be disregarded, too. As a result, only π_i and π_c apply, and the "same-soil" requirement can be dropped.

$$\pi_i = \frac{F}{\rho \ell^2 v^2}$$

$$\pi_c = \frac{F}{c \ell^2}$$

$$\pi_1 = \frac{\pi_i}{\pi_c}$$

$$\frac{c}{\rho v^2} = \frac{c'}{\rho' v'^2}$$

$$\frac{F}{c \ell^2} = \frac{F'}{c' \ell'^2}$$

$\left.\begin{array}{c}\\\\\\\end{array}\right\}$ model design rules for clay

Sandy soils are composed of relatively large particles having very little cohesion; hence, π_c can be neglected. Since the inner surface of sand is small (as compared to clay), weight forces are substantial and must be taken into account, along with internal friction between particles. Therefore, the pi-numbers of π_i, π_g, and π_μ apply.

$$\pi_i = \frac{F}{\rho \ell^2 v^2}$$

$$\pi_g = \frac{F}{\rho g \ell^3}$$

$$\pi_\mu = \mu$$

(Froude)

$$\pi_2 = \frac{\pi_i}{\pi_g} = \frac{g \ell}{v^2}$$

$g = g'$

(same soil)

$\rho = \rho'$

$$\frac{\ell}{v^2} = \frac{\ell'}{v'^2}$$

$$\frac{F}{\ell^3} = \frac{F'}{\ell'^3}$$

$$\mu = \mu'$$

$\left.\begin{array}{c}\\\\\\\\\end{array}\right\}$ model design rules for sand (same soil)

If the same sand is used for model and prototype, the grain size cannot be scaled. The effect of grain size, however, may be quite small since earth-working processes are not concerned with single grains but with the gross effect of a very large number of grains.

An effect that often cannot be neglected is that of soil compaction. Soil compaction, induced by the soil pressures and stresses of the deformation process, may considerably alter all soil values including density, cohesion, and internal friction, in a manner hard to predict. The only way to cope with this problem would be to use the same soil for the model as for the prototype and expose it to the same stresses and pressures. The "same-soil" and "same-stress" requirements are easily satisfied for clayey soils because here the scale factor of stress, F/ℓ^2, is unity for $c = c'$. But clayey soils usually develop very little compaction; hence, the "same-soil" requirement does not have to be imposed, and different clay types (with different cohesion) can be chosen for model and prototype.

By contrast, sandy soils are often easily compacted (or loosened) when subjected to pressures and stresses. Unfortunately, the pressures and stresses of the model are much smaller than those of the prototype. According to the developed model rule for sand,

$$\frac{F}{\ell^3} = \frac{F'}{\ell'^3} \longrightarrow \frac{p}{p'} = \frac{\ell}{\ell'} > 1$$

$$p \triangleq \frac{F}{\ell^2}$$

where p is the pressure or stress.

As a result, the model sand is less dense than the prototype sand (if the soil is compactible), in violation of the "same-sand" requirement. For originally dense sand, the effect of this violation will be small; for loose sand, it may be unacceptably large.

The two groups of model design rules, one for clay and one for sand, are only approximately valid for real soils, of course. Before any real soil is used for model tests, it should be pretested in simple model tests to find out whether it is "sand" or "clay;" if neither description fits, it is unsuitable for model tests.

5.1   Some Model Tests to Classify a Soil as Sand or Clay

In a presumably clayey soil, three geometrically similar rectangular blades
were moved slowly (slowly, to avoid inertial effects), one at a time, in a
horizontal direction through the same soil, with the width to penetration
ratio, t/z, kept constant. For all three plates, the bulldozing force F was
measured at steady-state condition. From the three forces, two ratios F/F'
were computed, with F' the force of either one of the two smaller plates,
and F the force of the largest plate, and plotted versus t/t' in a logarithmic
scale, Fig. 5.1. The plot revealed that F/F' was proportional to $(t/t')^2$.
Consequently, $F \propto t^2$ -- a relation predicted by the pi-number π_c for clay
with $c = c'$. Hence, the soil could be considered to be clay.

In a sandy, well compacted soil, three geometrically similar rectangular
plates were moved slowly (to avoid inertial effects) through the same soil in
a vertical direction. Here, the penetration force was not a constant as in the
previous test but changing with penetration depth, z . To compare forces in
geometrically similar situations, the ratio F/F' was plotted versus those
depth ratios, z/z', that were proportional to t/t' (the ratio of plate
dimensions). Figure 5.2 shows (in a logarithmic scale) that F/F' was proportional to $(t/t')^3$ or $(z/z')^3$. Consequently, $F \propto t^3$ -- a relation predicted
by the pi-number π_z for well compacted sand with $\rho = \rho'$. Hence, the soil
could be considered to be sand.

Source of Information

D.J. Schuring and R.I. Emori, "Soil deforming processes and dimensional
analysis," *SAE Trans.* 73, 485-494 (1965).

5.2   Off-Road Vehicle

From 1957 through 1960, when the U.S. Army was testing a number of large
off-road vehicles on different natural terrains, a few of the tests were
conducted to demonstrate the degree of possible correlation between model
and full-scale performance. Because the soil was judged to be of the sandy

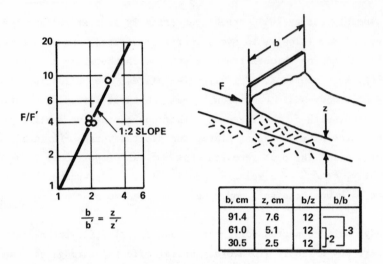

Fig. 5.1.   Bulldozing resistance F of three geometrically similar blades in clay soil with cohesion
c = 2.4 kN/m² and coefficient of internal friction μ = 0.05 ≈ 0

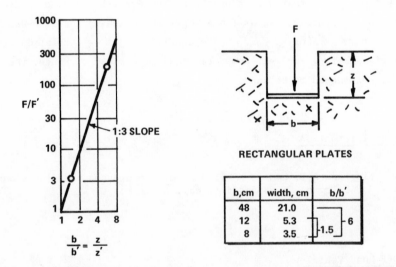

Fig. 5.2.   Penetration resistance F of three geometrically similar plates into sand with cohesion
c ≈ 0 and coefficient of internal friction μ = 0.34.

type, model and prototype tests were performed at the same sites on the same
soil. However, since natural soil is never homogeneous, but always stratified and changing its properties with changes in moisture content, only
approximate agreement between model and full-scale tests was expected.

The tests were performed with a high flotation vehicle (dubbed Marsh Buggy,
see Fig. 14, Part I) and its 1/4 scale model, equipped with four huge
balloon tires, to keep the soil pressure low. The soil, then, could be
considered loaded by four flexible spherical shells that would exert vertical and horizontal forces -- a situation involving modeling not only of soil
properties but also of the deformation characteristics of the tires. Since
deformations took place at low speeds, the pi-number applicable to soil
modeling was π_g , provided the "same-soil" approach was used and compaction
effects were neglected. Then, $F/\ell^3 = F'/\ell'^3$. The same rule was applied to
modeling the tires, with ℓ representing not only the tire dimensions, but
also the tire deflections under load, so that $\delta/\delta' = d/d' = \sqrt[3]{F/F'}$, where
δ is wheel deflection, and d is tire section height. (Of course, any other
characteristic tire dimension would have been just as adequate.)

The full-scale Marsh Buggy and its 1/4-scale model were tested at various
gross weights. At each gross weight, three relative tire deflections were
realized by altering the internal tire pressure -- δ/d = 8%, 17%, and 30%, as
measured on a hard surface. The test course of sandy soil was tilled before-
hand to reduce inhomogeneities. A test consisted of running the vehicles
at low speed along the test course and measuring the drawbar pull, B, and the
wheel sinkage, $\jmath$, at discrete increments of slip between 0% and 70%.

Drawbar pull, i.e., the capability of a vehicle to pull a horizontal load
(say, a trailer) at constant speed, is a strong function of slip, with slip
defined as the speed of the free-rolling vehicle (moving on a hard surface)
minus the actual vehicle speed, and this speed difference made dimensionless
by the free-rolling speed. Figure 5.3 shows a typical drawbar pull-slip
curve, with a sharp increase of drawbar pull at low slip values and a slight
decline at higher slip values. Often, a distinct "hump" is observed at
around 20% slip.

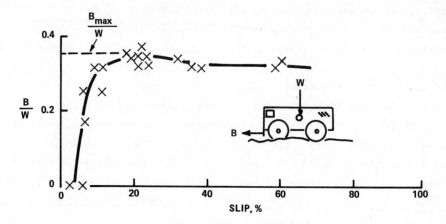

Fig. 5.3.   Typical drawbar test data in sand.

B = DRAWBAR PULL
W = VEHICLE WEIGHT

At the tests, changing slip was implemented by changing the drawbar pull. Since slip is a dimensionless number it should become the same for both model and prototype when they are operated at the maximum value of drawbar pull. This was indeed the case; the maximum drawbar pull was observed at approximately 20% for both vehicles.

Test results are shown in Figs. 5.4 and 5.5, with both the maximum drawbar coefficient, B_{max}/W, and the relative wheel sinkage, z/D, plotted vs the vehicle gross weight, W (in prototype scale). The correlation between the model's and prototype's maximum drawbar pull coefficients was good for all three relative tire deflections at all gross weights. The correlation of sinkage data was less satisfactory, mainly because of the difficulty of measuring the wheel's true sinkage which, during the wheel's passage through sand, was obscured by wheel deformation, and after passage, by sand inflow into the rut.

Source of Information

U.S. Army Transportation Res. Command, *Scaled Vehicle Mobility Factors (Sand)* by Wilson-Nuttall-Raymond, TR-61-67, Fort Eustis, Va., Apr 1961.

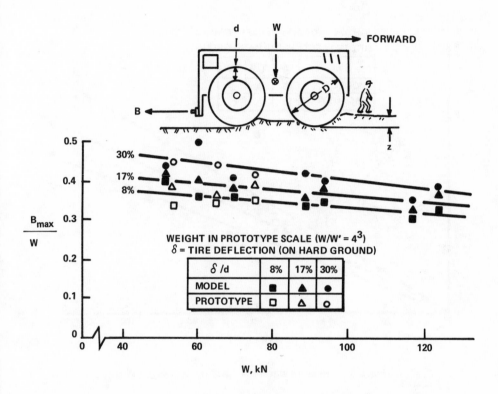

Fig. 5.4.   Maximum drawbar pull coefficient, B_{max}/W, vs. weight, W.

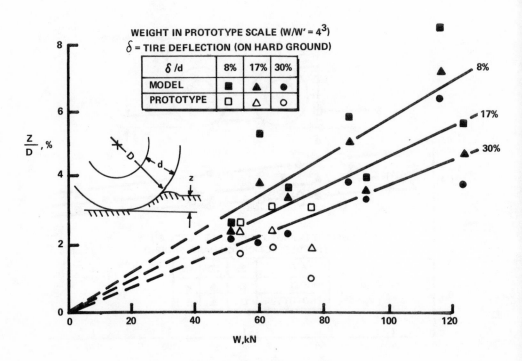

Fig. 5.5.   Relative wheel sinkage, Z/D, vs. weight, W.

5.3 Howitzer Foundation -- Dynamic Response

This example involves inertial forces of the soil as well as of the engaged equipment.

Each time a field artillery piece is fired when it is positioned on soft soil, it is slightly but permanently displaced, so that its firing accuracy is impaired with each shot. Rotational displacement was found to more damaging to the weapon accuracy than linear displacement. To design a howitzer that would shift linearly and would not rotate during firing, weapon designers needed to analyze the dynamic interaction between weapon foundation and soil. Since the necessary data were lacking, model tests were set up to study the dynamic response of sand and clay to artillery firing.

The major elements of a howitzer are the barrel, the recoil mechanism, and the foundation. The recoil mechanism takes up the initial impulse of the discharged projectile and transmits it to the foundation within a few hundred milliseconds. For a howitzer of the considered class (towed, 105 mm, M2A2), the impulse exerted onto the foundation is essentially of constant magnitude, with the force ranging up to 200 kN, and the force duration up to 600 ms. The foundation usually consists of the two wheels and two stakes with attached spades, which provide the anchoring forces (Fig. 5.6).

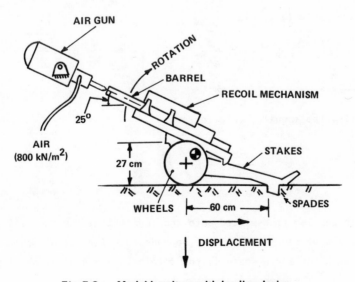

Fig. 5.6.   Model howitzer with loading device.

The problem was to design a howitzer model that would duplicate in scaled fashion the dynamic response of the prototype. The dynamic response would depend on the impulse exerted onto the foundation, on the inertial and geometrical properties of the howitzer, and of course on the soil's reaction to the transmitted forces. The soil's reaction forces would be composed of inertial forces and static forces (cohesion forces, for clay; internal friction and weight forces, for sand). Consequently, the two sets of model rules, one for clay and one for sand, would apply fully.

For clay, the force scale factor is, with $c = c'$, and $\rho = \rho'$,

$$F/F' = (\ell/\ell')^2$$

and the time scale factor,

$$v^2 = v'^2 \qquad \longrightarrow \qquad t/t' = \ell/\ell'$$
$$v \triangleq \frac{\ell}{t}$$

Therefore, if the same clay is used for model and prototype, the impulse scale factor is

$$\frac{Ft}{F't'} = \left(\frac{\ell}{\ell'}\right)^3$$

For sand, the force scale factor is (same soil for model and prototype)

$$F/F' = (\ell/\ell')^3$$

and the time scale factor is

$$\frac{\ell}{v^2} = \frac{\ell'}{v'^2} \qquad \longrightarrow \qquad \frac{t}{t'} = \sqrt{\frac{\ell}{\ell'}}$$
$$v \triangleq \frac{\ell}{t}$$

Consequently, the impulse scale factor is

$$\frac{Ft}{F't'} = \left(\frac{\ell}{\ell'}\right)^{3.5}$$

Besides the impulse transmitted onto the foundation, the inertial properties and the weight of the howitzer must be modeled. The weight scale factor is identical with the force scale factor developed for scaling the impulse for sand and clay. The moment of inertia scales as follows:

$$I \triangleq m\ell^2 \longrightarrow I \triangleq \rho\ell^5 \longrightarrow \frac{I}{I'} = \left(\frac{\ell}{\ell'}\right)^5$$

$$m \triangleq \rho\ell^3 \quad\quad\quad \rho = \rho'$$

The developed scale factor for the moments of inertia is based on the equality
of the model's and prototype's density. As a consequence, the weight scale
factor is $W/W' = m/m' = \left(\ell/\ell'\right)^3$, in agreement with the force scale factor for
sand but, unfortunately, in violation of the force scale factor for clay. It
could be argued that weight forces, i.e., static forces imposed by the howitzer onto the soil, are insignificant in view of the expected high dynamic
forces, but still the impact of the weight neglection on the experimental
results was uncertain.

The experiments were performed with a 105-mm M2A2 towed howitzer and its
1/5-scale model in sand and clay. For the prototype tests, rounds were fired
using water projectiles with a propelling charge of "zone 5" for sand and
of "zone 7" for clay, at an elevation angle of $25°$ to pass the load directly
through the two spades. The wheels were not locked. A series of five rounds
was repeated three times on clay and sand, each time at a different location,
with rotations and linear displacements of the gun measured after each shot.

A schematic drawing of the model howitzer is shown in Fig. 5.6. A square-
wave impulse was applied to the foundation with a specially designed air
gun that could generate, in controlled fashion, forces between 100 and 5000 N
at durations between 3 and 100 ms. From the "zone 5" charge, a model force
of 249 N acting for 76 ms was computed and applied on sand; and from the
"zone 7" charge, a model force of 1870 N acting for 37 ms on clay. The
actual prototype data were not given in the Source of Reference. Table 5.1
shows the prototype data as computed from the model rules derived earlier.
Accordingly, the prototype impulse force and duration must have been 31 kN
and 170 ms for sand (zone 5), and 47 kN and 185 ms for clay (zone 7).

The model weight was made 124 N for both sand and clay. Actually, with a
prototype weight of 15.5 kN, the model weight should have been 620 N for
clay (i.e., 5 times higher than the weight for sand), but the inertial forces

S.M.E.—G

TABLE 5.1. Some Test Data for Howitzer and Model. Prototype Data Re-Computed from Model Data Given in Source of Information.

| SOIL | SQUARE-WAVE LOADING | | | | | | WEIGHT, N | | | MOMENT OF INERTIA AROUND C.G., kg m^2 | | |
| | FORCE, N | | | TIME, ms | | | | | | | | |
	SCALE FACTOR	MODEL	PROTOTYPE^2	SCALE FACTOR	MODEL	PROTOTYPE^2	SCALE FACTOR	MODEL	PROTOTYPE^2	SCALE FACTOR	MODEL	PROTOTYPE^2
SAND	5^3	249	31×10^3	$\sqrt{5}$	76	170	5^3	124	15.5×10^3	5^5	0.213	665
CLAY	5^2	1870	47×10^3	5	37	185	5^2	124 [1]	15.5×10^3	5^5	0.213	665

[1]IN VIOLATION OF THE CORRECT MODEL RULE OF $W' = W/5^2 = 620$ N.

[2]NOT GIVEN IN SOURCE OF INFORMATION. HERE, COMPUTED FROM MODEL RULES.

would then have been much too high. Therefore, the weight influence was tentatively discounted in favor of correct scaling of the moment of inertia, as indicated in Table 5.1.

The model tests were performed similarly to the prototype tests. A series of five shots was repeated four times on clay and again on sand, each time at a slightly different location, with rotation and linear displacements of the foundation measured after each shot. Figure 5.7 depicts the horizontal displacements and rotations of model and prototype in sand after one to six shots. Figure 5.8 shows the same relations when the pieces were mounted on clay. All model data are plotted in prototype scale.

In sand, the large scatter of model data defeats precise interpretation. Hence, the influence of violating the "equal-stress" condition could not be investigated. Very probably, its influence was small as compared to the influence of other factors such as soil non-uniformities which presumably caused the large data scatter. Sand data are always subject to large scatter unless extreme precautions are taken in soil preparation. The scatter can only be offset by a very large number of test repetitions, preferably twenty or more. Nevertheless, even with only four repetitions, the

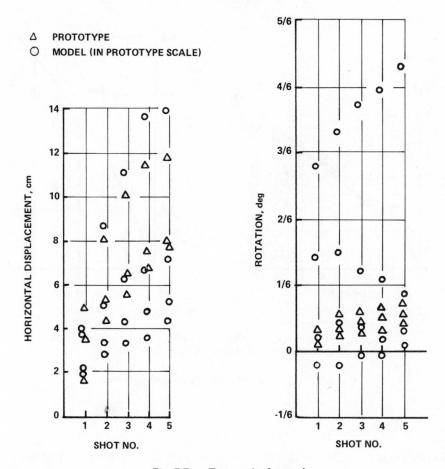

Fig. 5.7.   Test results for sand.

average model data of the horizontal displacement are fairly close to the
average prototype data. The scatter of the model's rotation data is too
large, however, to draw valid conclusions. Note though that the rotations
did not exceed one (1) arc degree and were in most cases less than ten arc
minutes. These small rotations are presumably very sensitive to soil
inhomogeneities.

In clay, the data exhibited very little scatter. Unfortunately, rain
modified the water content of the clay after the prototype and before the
model tests. Model tests had to be performed at three different water contents, and the data for the water content of the prototype soil (5.1%)

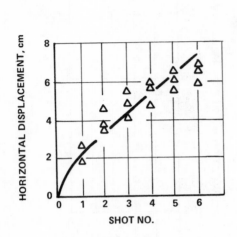

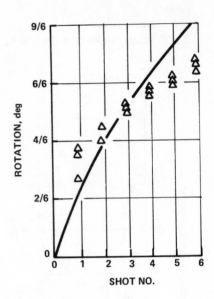

△ PROTOTYPE DATA

—— MODEL CURVE, INTERPOLATED FROM RESULTS OF MODEL TESTS ON
CLAY WITH WATER CONTENTS OF 5.1%, 10.6%, and 28.6%

Fig. 5.8.   Test results for clay (5.1% water content).

produced by graphical interpolation. The thus established model data plotted
in prototype scale agree well with the prototype data despite neglection of
the influence of the weapon's weight.

Sources of Information

Southwest Res. Inst., *Modeling Studies of the Response of Weapon Foundation
in Soils*, Phase II, Final Report, by P.S. Westine, SwRI Proj. 02-1548, San
Antonio, Tex., Mar 1966.

U.S. Army Engr. Waterways Experiment Station, *Theoretical and Experimental
Studies on Dynamic Response of Weapon Foundations* by P.S. Westine, Contract
Rep. No. 3-173, Vicksburg, Miss., Oct 1967.

Related References

Earthmoving Equipment

K.K. Barnes, et al, "Similitude in studies of tillage implement forces," *Agricultural Eng.* 41, 1, 32-37,42 (Jan 1960).

D.E. Cobb, et al, "Scale model evaluation of earthmoving tools," *Proc. First Intl. Conf. on Mechanics of Soil-Vehicle Systems*, Edizioni Minerva Technica, Torino, Italy, 1962, pp. 412-428.

U.S. Army Tank Autom. Ctr., Components R&D Labs., *Dimensional Analysis of Load-Sinkage Relationships in Soils and Snow* by R.A. Liston and E. Hegedus, TR-8692 (LL100), Warren, Mich., Dec 1964.

R.J. Sullivan, "Earthmoving in miniature," *J. Terramechanics* 1, 4, 85-106 (1964).

E. Hegedus, "Plate sinkage study by means of dimensional analysis," *J. Terramechanics* 2, 2, 25-32 (1965); discussion in 4, 2, 59-64 (1967).

K. Baganz, "Untersuchungen ueber Modellbeziehungen bei Bodenbearbeitungs-werkzeugen" (Model investigations of tillage tools), *Deutsche Agrartechnik* 15, 12, 555-558 (Dec 1965).

D.F. Young, "Similitude of soil-machine systems," *J. Terramechanics* 3,2,57-70 (1966).

G.T. Cohron, "Model testing of earthmoving equipment," *Trans. ASAE* 9,2, (1966).

R.A. Liston, "Dimensional analysis in land locomotion problems," *Res. Rep. No. 6* by staff of Land Locomotion Lab., TR-9560, U.S. Army Tank Autom. Ctr., Components R&D Labs., Warren, Mich., Nov 1966, pp. 167-208.

B.W. Firth, "Resistance of soils to sinkage and translation of rigid bodies: a study by means of dimensional analysis," Paper 67012 presented at SAE Autom. Eng. Congr., Detroit, Mich., Jan 1967.

U.S. Army Tank Autom. Command, Mobility Systems Lab., *Dimensional Analysis of the Pressure-sinkage Relationship in Non-clay Mineral Materials* by Bong-Sing Chang, TR-10006 (LL 122), Warren, Mich., Jun 1967.

R.J. Garrity, et al., "A test comparison of model and full-size bulldozer blades," Paper 680612 presented at SAE Natl. Combined Farm Construction and Industrial Machinery, Powerplant, and Transportation Meetings, Milwaukee, Wisc., Sep 1968.

C.A. Reaves, et al., "Similitude in performance studies of soil-chisel systems," *Trans. ASAE* 11, 5, 658-660, 664 (Sept-Oct 1968).

R.L. Schafer, et al., "Model-prototype studies of tillage implements," ibid., 661-664.

L.W. Larson, et al., "Predicting draft forces using model moldboard plows in agricultural soils," ibid., 665-668.

K.H. Roscoe, "Soils and model tests," *J. Strain Analysis* 3, 1, 67-74 (1968).

A. Soltynski, "Physical similarity and scale effects in soil-machine systems," *J. Terramechanics* 5, 2, 31-43 (1968).

P.A. Hustad and W.R. Cox, "Force-penetration characteristics of a sand horizontally penetrated by plates, cones, and spherical segments," NASA CR-1252, Washington, D.C., Jan 1969.

J.T. Gray, "Soil bin scale-model testing," Paper 690357 presented at SAE Earthmoving Industry Conf., Central Illinois Sect., Peoria, Ill, Apr 1969.

D.R. Freitag, et al., "Similitude studies of soil-machine systems," *J. Terramechanics* 7, 2, 25-59 (1970).

V.M. Matsepuro, "Study of the resistance of the soil and ground by methods of the theory of similitude," Transl. from Russian by W.R. Gill; U.S. Dept. Agriculture, Natl. Tillage Machinery Lab., Auburn, Ala., Oct 1972.

R.L. Schafer, D.R. Freitag, and R.D. Wismer, "Distortion in the similitude of soil-machine systems," *J. Terramechanics* 9, 2, 33-64 (1972).

R.D. Wismer, D.R. Freitag, and R.L. Schafer, "Application of similitude to soil-machine systems," Paper presented at 6th Seminar on the Similitude of Soil Machine Systems at U.S. Dept. Agriculture, Natl. Tillage Machinery Lab., Auburn, Ala., Feb 1975 (publ. by ASAE).

Off-Road Vehicles

Canadian Armament R&D Establishment, *A Method of Evaluating Scaling Factors and Performance of Tracked Vehicles on Snow* by W.J. Dickson, et al., CARDE Tech. Memo 612/61, Valcartier, Quebec, Jun 1961.

C.J. Nuttall and R.P. McGowan, "Predicting equipment performance in soils from scale model tests," Paper 408A presented at SAE Heavy Duty Vehicle Meeting, Milwaukee, Wisc., Sep 1961.

H.H. Hicks, "A similitude study of the drag and sinkage of wheels using a system of soil values related to locomotion," *Proc. First Intl. Conf. Mechanics of Soil-Vehicle Systems*, Edizioni Minerva Technica, Torino, Italy, 1962, pp. 645-655.

C.J. Nuttall and R.P. McGowan, "Scale models of vehicles in soils and snow," ibid., pp. 656-677.

W.L. Harrison, "Analytical prediction of performance for full-size and small-scale model vehicles," ibid., pp. 678-702.

Canadian Armament R&D Establishment, *Variations in the Performance of a Tracked Model Vehicle on Loose Sand Due to Changes in the Longitudinal Position of its Center of Gravity* by D.A. Nicholson and G.E. Booker, CARDE Tech. Memo 715, Valcartier, Quebec, Jan 1963.

U.S. Army Tank Automotive Ctr., Components R&D Labs., *Rigid Wheel Studies by Means of Dimensional Analysis* by E.T. Vincent, et al., TR-7841, Warren, Mich., Apr 1963.

J.M. Clark, H.P. Simon and C. Roma, "Correlation of prototype and scale model vehicle performance in clay soils," *SAE Trans.* 73, 272-294 (1965).

D.R. Freitag, "A dimensional analysis of the performance of pneumatic tires on clay," *J. Terramechanics* 3, 3, 51-68 (1966).

I.R. Ehrlich, "The place of model tests in vehicle development," Paper 670169 presented at SAE Automotive Eng. Congr., Detroit, Mich., Jan 1967.

V.C. Pierrot and W.F. Buchele, "A similitude of an unpowered pneumatic tire," *Trans. ASAE* 11, 4, 673-676 (July - Aug 1968).

U.S. Army Engineer Waterways Experiment Station, *Performance of Soils under Track Loads, Rep 2* (Prediction of track pull performance in a desert sand), by G.W. Turnage, TR-M-71-5, Vicksburg, Miss., Nov 1971.

B.W. Firth, "Characteristic lengths in soil-vehicle mechanics," *J. The Franklin Inst.* 292, 6, 491-498 (Dec 1971).

G.D. Swanson, "Scale-model tire tests in clay," *J. Terramechanics* 10, 3, 21-27 (1973).

High-Speed Deformation of Soil

D.R. Reichmuth, *Soil-Projectile Interaction during Impact*, Univ. Texas, Dept. Civil Eng., Ph.D. Diss., Austin, Tex., Jan 1967.

K. Awoshika and W.R. Cox, "An application of similitude to model design of a soil-projectile system," NASA CR-1210, Washington, D.C., Nov 1968.

D. Schuring, "A contribution to soil dynamics," *J. Terramechanics* 5, 1, 31-37 (1968).

W.E. Baker, P.S. Westine, and F.T. Dodge, "Modeling in soil dynamics," Ch. 11, *Similarity Methods in Engineering Dynamics*, Hayden, Rochelle Park, N.J., 1973.

CASE STUDY 6

Equilibrium Temperature of a Large Tire

During the past ten years or so, great analytical and experimental efforts
have been made to predict tire performance. However, despite considerable
progress in tire theory, a fully mathematical model is not yet in sight,
and tires are still developed very much empirically. This is acceptable for
automobile tire development where large test facilities are not required and
initial development costs are low in comparison with the cost of the ultimate
production run. Large earthmover tire development, however, requires large
test facilities, and the tires are sold in small quantities. Thus, the idea
of testing small replicas instead of big tires is attractive.

A tire must meet a wide spectrum of requirements such as load carrying, road
holding, shock cushioning, and -- on natural terrain -- obstacle enveloping
and flotation. Equally important are wear and cut resistance and temperature
buildup. It is impossible to model all these functions with just one model.
As is true in most fields of model application, a series of partial models
must be used, each covering a restricted set of functions. The following
study was made to predict the equilibrium temperature of a large earthmover
tire from a small model.

In every mechanism, some of the input energy is lost. Temperature equilibrium
is achieved if the loss energy is dissipated at constant rate. To analyze
the loss energy of a tire at the state of temperature equilibrium, we assume
that the ground is rigid so that no energy is needed to deform the ground.
We can then divide the loss into two major parts. One part is originated in
the bulk of the tire, the other one in the immediate vicinity of the contact
area between tire and ground. The bulk deformation loss is caused by gross
tire flexure, that is, by adaptation of the circular tire to the flat ground
and to large obstacles.

The contact area losses can be further divided into three parts: abrasional
losses, adhesional losses, and micro-deformational losses. Since abrasion

181

losses are small -- less than one percent of the total losses -- we can neglect abrasion. Adhesional losses are attributed to intermolecular forces
between rubber and ground; micro-deformational losses, to the rapid deformation of the tread rubber by small ground asperities. Both these losses are
also proved to be rather small.

The energy generated in the tire and between tire and ground is stored in
the tire and carried away through (a) conduction within the tire and into the
ground; (b) convection from the tire surface into the surrounding air; and
(c) radiation. Because the surface temperature of a tire is rather low, the
amount of heat dissipated through radiation is small and can be neglected.
Furthermore, since we are concerned with steady-state conditions, stored
heat would remain unaffected and need not be considered.

Governing Laws. As a result, we are left with three modes of energy loss at
steady-state: (1) heat energy generated in the tire bulk through gross tire
flexure; (2) heat energy conducted within the tire and into the ground; and
(3) heat energy dissipated through convection from the tire into the air.
We will now discuss the requirements for modeling these energies.

(1)   The general nature of heat generated by flexure of rubber is fairly
      well understood. Almost all authors attribute deformation heat to
      visco-elastic properties of the rubber. A simple example of a visco-
      elastic body is presented by the Kelvin-Voigt body consisting of a
      spring and a viscous damper. If the body is sinusoidally flexed, the
      work dissipated per unit cycle is

$$Q = \pi \sigma \epsilon \; \frac{\tan \delta}{\sqrt{1 + \tan^2 \delta}} \cdot V$$

      where ϵ is strain amplitude, σ is stress amplitude, V is flexed volume,
      and δ is the phase angle between ϵ and σ. For the Kelvin-Voight body,
      the phase angle is

$$\tan \delta = \frac{\omega k}{c}$$

where $tan\ \delta$ is called the loss tangent, ω is the radian frequency, c is the spring rate, and k is the damping rate.

For rubber and rubberlike materials, energy dissipation due to flexure follows a less simple course. However, we infer from dimensional reasoning that the work per cycle necessary to deform a bulk element (even if it is interspersed with cords) can be expressed by a relation constructed like the one for a Kelvin-Voight body:

$$Q \triangleq \sigma \epsilon\, VM(f, \theta, \sigma) \quad \text{or} \quad \pi_B = \frac{Q}{\sigma \epsilon\, V M}$$

where M is a function reflecting the visco-elastic properties of the cord-strengthened rubber. The function M is assumed to be dependent on the temperature, θ, the deformation frequency f, and the average stress, σ.

(2) Heat conduction is governed by Fourier's law of energy transfer. Applied to a small cross-sectional element, it can be expressed as

$$d\dot{Q} = k\,(\partial\theta/\partial n)\,dA$$

where $d\dot{Q}$ is the rate of heat flowing through a small control area dA located in any direction within the body, $\partial\theta/\partial n$ is the temperature gradient in direction n normal to the control area, and k is the thermal conductivity of the material.

Expressed in representative terms, Fourier's law assumes the form

$$Q \triangleq k\,\frac{\Delta\theta}{\ell}\,\ell^2 t \longrightarrow \pi_k = \frac{Q}{k\,\Delta\theta\,\ell t}$$

where t is time.

(3) The amount of heat energy removed from the tire by the surrounding cooler air is formally described by Newton's equation

$$d\dot{Q} = h\,\Delta\theta\,dA$$

where $d\dot{Q}$ is the rate of heat energy passing through a small control area dA at the surface of the tire, $\Delta\theta$ is the temperature difference between tire and air measured at defined locations outside the heat

boundary layer, and h is the surface coefficient of heat transfer
between tire and air. Expressing this equation in representative
terms, we obtain

$$Q \triangleq h \Delta \theta \ell^2 t \longrightarrow \pi_h = \frac{Q}{h \Delta \theta \ell^2 t}$$

Note that the coefficient h is not a material property but, similar
to M, a function of many variables such as air properties, velocity,
and surface roughness.

Modeling M and h. A quantitive description of the function $M(f, \theta, \sigma)$
is lacking. Now, if insight into a process that must be physically modeled
cannot be achieved, as is true here, the "same material" approach often offers
a solution: although we don't know the way the material properties vary with
frequency, pressure, temperature, etc., we know at least *what* quantities are
governing the material's behavior. Consequently, when using the same material
for both prototype and model, we must expose the model to the same frequencies,
temperatures, and stresses that are governing the prototype, thus forcing
the model material to respond in the same way as the prototype material does.
In other words, the scale factors of frequency, temperature, and stress must
all be kept unity.

To relax these rather stringent requirements, we will investigate how strongly the material property M depends on f, θ, and σ. Since the function M
represents energy loss in a way analogous to the loss tangent, and since the
loss tangent of rubber and rubberlike materials is a well understood property,
whereas the tire property M is not, we will use the loss tangent as a substitute for M, assuming that they both exhibit analogous trends. The analogy
is assumed to hold even for rubber that is reinforced by cords.

For natural rubber which has low hysteresis at room temperature ($20^{\circ}C$), the
loss tangent is nearly independent of deformation frequency up to roughly
100 Hz. Then, with increasing frequency, the loss tangent increases until
it passes through a maximum. Loss tangent curves shown in Fig. 6.1a suggest
that this maximum occurs at frequencies much higher than 10^4 Hz. By contrast, butyl, a high hysteresis rubber, exhibits a maximum loss tangent at
relatively low frequencies, Fig. 6.1b.

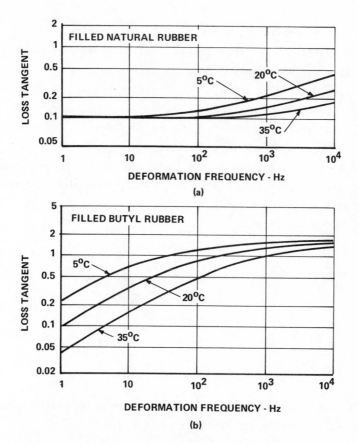

Fig. 6.1.   Frequency and temperature dependence of loss tangent; (a) low hysteresis rubber,
(b) high hysteresis rubber; adapted from J.C. Snowdon (see Ref.).

The loss tangent is also a function of temperature. Higher temperature causes
the loss tangent-frequency curve to shift toward higher frequencies for both
low hysteresis and high hysteresis rubberlike materials. However, since at
low frequencies the loss tangent of natural rubber runs nearly parallel to
the frequency axis, a shift along the frequency axis does not change the loss
tangent. Hence, at room temperature, it can be considered independent of
frequency and temperature up to at least 100Hz. At higher temperature (35°C),
the loss tangent may be constant up to 1000 Hz, Fig. 6.1a. For butyl rubber,
however, it is a function of temperature at all frequencies.

Large earthmover tires are usually made from natural rubber. Also, they
are ordinarily run at temperatures not exceeding 100°C. Hence, we may
assume that the loss tangent and, along with it, the rubber property M,

is independent of temperature and time up to approximately 1000 Hz, as
Fig. 6.1a suggests. The maximum frequency of bulk deformation of an
earthmover tire can be estimated from the average distance of (terrain)
obstacles, a, and the tire speed, v. If, for instance, a = 5 cm and
v = 50 km/h, then f = 280 Hz (that is, < 1000 Hz). Consequently, for earthmover tires, the material property M can be considered a function of average
stress only. We do not know this function but we can minimize its influence
on M by using the same low-hysteresis rubber for model and prototype and by
keeping the stress scale factor at unity. Then, the model rule ensuing
from π_B assumes a very simple form:

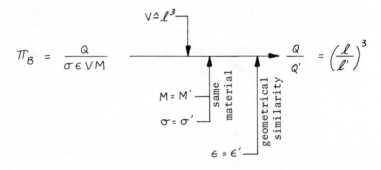

$$\pi_B = \frac{Q}{\sigma \epsilon V M}$$

$$V \triangleq \ell^3$$

$$M = M' \quad \sigma = \sigma' \quad \epsilon = \epsilon'$$

$$\frac{Q}{Q'} = \left(\frac{\ell}{\ell'}\right)^3$$

With the "same material" approach, then, the heat scale factor turns out to
be proportional to the length scale factor cubed.

The heat transfer coefficient h cannot be maintained the same for model and
prototype, as follows from a comparison of π_k and π_h .

$$\pi_k = \qquad \xrightarrow{\text{same material} \atop k = k'} \quad \frac{Q}{Q'} = \frac{\Delta \theta \ell t}{\Delta \theta' \ell' t'}$$

$$\pi_h = \frac{Q}{h \Delta \theta \ell^2 t} \qquad \xrightarrow{} \qquad \frac{Q}{Q'} = \frac{h \Delta \theta \ell^2 t}{h' \Delta \theta' \ell'^2 t'}$$

$$\frac{h}{h'} = \sim$$

Naturally, the length scale factor ℓ/ℓ' is selected larger than unity so
that the model would be smaller than the prototype. Consequently, the model's
coefficient of heat transfer has to be ℓ/ℓ'-times larger than that of the
prototype. The coefficient of heat transfer can be increased by increasing

the air speed; for instance, by installing a large fan in front of the model and by selecting an air speed that would satisfy the model rule for h.

Modeling Stresses. A tire is an inflated shell composed of a multilayered network of cords (so-called plies), embedded in rubber, as indicated in Fig. 6.2. Stressed under inflation pressure and external loads, the flexible but inelastic cords change their relative position slightly, thereby stressing and deforming the soft and elastic rubber matrix and, thus, generating heat. Consequently, proper modeling of bulk heat presupposes proper modeling of cord structure and arrangement.

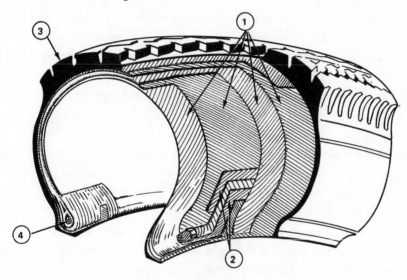

Fig. 6.2.   Conventional bias-ply tire.

(1) CARCASS PLIES
(2) BEAD REINFORCEMENT
(3) TREAD
(4) BEAD

Cords are made of thin fibers twisted into strands that are then twisted together again, a complex structure that is expensive and cumbersome to scale. In an attempt to circumvent this difficulty, Tsukerberg and Gordon (see References) replaced "microscopic" modeling of the cords and their arrangement by "macroscopic" modeling. They argued that instead of faithfully scaling the cord structure, their distance, and their arrangement in plies, model fabrication would be much more economical if one could use the

same fabric built into the prototpye. Of course, only a $(\ell'/\ell)^2$-th part of
the original fabric would then be moldable into the model; and the total
cross-sectional cord force, F', of the model would relate to that of the
prototype, F, as

$$\frac{F}{F'} = \left(\frac{\ell}{\ell'}\right)^2$$

This relation is equivalent to the equal-stress requirement. Formally,

$$\frac{\sigma}{\sigma'} = \frac{F/\ell^2}{F'/\ell'^2}$$

where σ represents any stress including inflation pressure, pressures and
stresses within the tire footprint and within the tire carcass. For $\sigma = \sigma'$,
we have the same relation as derived from macroscopic modeling. Hence,
using the same cord fabric for model and prototype ensures maintaining the
same stresses for both -- a requirement that, incidentally, is in compliance
with the equal-stress requirement derived earlier for modeling the rubber
property M.

Tsukerberg and Gordon and later Clark and coworkers (see References) checked
experimentally the validity of these considerations by statically loading a
1/7 scale model tire equipped with full-scale cords and plies with a load
1/49th that of the prototype. The investigators measured the deflections
and contact areas of model and prototype at various loads and internal
pressures and obtained excellent agreement. Figure 6.3 presents an example;
the good agreement is remarkable. Usually, agreement between model and prototype lessens with increasing length scale factor, particularly if distortion
is involved. If we assume that Tsukerberg and Gordon's prototype tire had
14 plies, the model tire had only 2. One would expect a rather substantial
error at this large deviation from ideal similarity, which would require 14
plies in the model.

Tests. The good static results encouraged the use of a model with prototype
cords for heat modeling. The same cord fabric was used for both a 29.5-35
earthmover tire and its 1/4 scale model. Both tires were inflated to the
same pressure (about 350 kN/m^2) and then run, under load, at various constant
speeds, the prototype on a proving ground in Texas (for about 8 h per test),
the model on a drum in a laboratory (for about 2 h per test). During each
test, the temperatures of the road and drum surfaces and of the tire shoulders

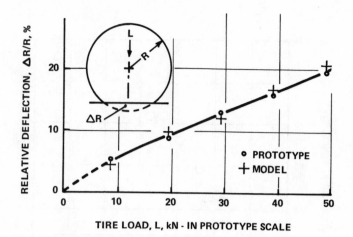

Fig. 6.3.   Static load—deflection curves of prototype and model tire, adapted from S.M. Tsukerberg and R.K. Gordon (see Ref.); length scale factor = 7.

were measured at short intervals and recorded. From the temperature-time plots, the equilibrium temperature of the tire shoulders, the road, and the drum were estimated. Finally, the difference between the equilibrium tempera- tures of the tire and the road, $\Delta\theta$, and of the model tire and the drum, $\Delta\theta'$, were plotted as a function of tire speed. Figure 6.4 shows the result.

From the earlier derived model rules, one would expect the following result:

$$\pi_B \longrightarrow \frac{Q}{Q'} = \left(\frac{\ell}{\ell'}\right)^3$$

$$\pi_k \longrightarrow \frac{Q}{Q'} = \frac{\Delta\theta\,\ell\,t}{\Delta\theta'\,\ell'\,t'} \longrightarrow \frac{\Delta\theta}{\Delta\theta'} = \frac{\ell}{\ell'}\,\frac{v}{v'}$$

$$t \triangleq \frac{\ell}{v}$$

$$\pi_h \longrightarrow \frac{h}{h'} = \frac{\ell'}{\ell}$$

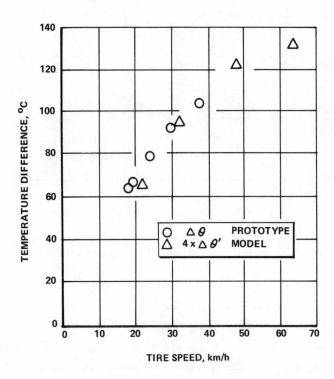

TIRE SPEED, km/h

○ 29.5 - 35 FULL SIZE EARTHMOVER TIRE, LOAD = 118 kN
✕ 1/4 SCALE MODEL; LOAD (EFFECTIVE)[1] = $118/4^2$ = 7.4 kN
$\Delta \theta$ = PROTOTYPE TEMPERATURE DIFFERENCE BETWEEN TIRE SHOULDER AND ROAD
$\Delta \theta'$ = MODEL TEMPERATURE DIFFERENCE BETWEEN TIRE SHOULDER AND ROAD

[1]THE MODEL TIRE WAS LOADED WITH A "RATED" LOAD OF 5.9 kN WHICH, ON THE
CURVED SURFACE OF A DRUM, INDUCES THE SAME DEFORMATIONS AS 7.4 kN ON
A FLAT ROAD.

Fig. 6.4.   Equilibrium shoulder temperature of model and prototype tires, as function of tire
speed.

Provided, then, that the heat transfer factor is modeled properly, the
temperature ratio $\Delta\theta\big/\Delta\theta'$ should assume the value of 4 at equal speeds of
model and prototype. Figure 6.4 shows that this is nearly true although h
was not modeled. Thus, the influence of heat convection at the given tire
speeds must have been small.

At lower speeds, the equilibrium temperature of both tires increases nearly
linearly with speed. This is in agreement with the model rule derived from
π_B and π_k which states that for unit length scale factor the equilibrium
temperature should increase linearly with speed. At higher speeds, temperature does not increase as fast as speed. Perhaps, with increasing temperature and, hence, increasing inflation pressure, the tire stiffens, thus
decreasing tire flexure and, with it, heat generation and temperature.

The equal-stress requirement includes equal inflation pressures of model
and prototype. Ideally, the tires should be inflated such that equal pressures are ensured at the end of a test at equilibrium temperature. However,
the tests were run under equal initial (cold) inflation pressures. The
impact of this inaccuracy on the test results could not be assessed.

Under these circumstances, it could not be clearly demonstrated that cord
distortion would not affect similarity in bulk heat although, judged by the
good test results, the effect of distortion appears to be rather small.

Sources of Information

D.J. Schuring, "Scale modeling of equilibrium tire temperature," *Tire
Science and Technology* 1, 3, 267-289 (1973).

S.M. Tsukerberg and R.K. Gordon, "Determination of the operating qualities
of automobile tyres by the modeling method," Royal Aircraft Establishment,
Ministry of Aviation, Library Transl. 1198, Nov 1966.

J.C. Snowdon, "Rubberlike materials, their internal damping role in vibration isolation," *J. Sound and Vibration* 2, 175-193 (1965).

S.K. Clark, et al, "Structural modeling of aircraft tires," *J. Aircraft*
9, 2, 162-167 (Feb 1972). Also: NASA-CR-2220, Mar 1973.

CASE STUDY 7

Free-Burning Fires

Ancient and modern civilization would be unthinkable without the controlled
use of free-burning fires; yet, it is only in recent years that fire research
has received strong attention.

A free-burning fire can be considered a cyclic process. Some of the
generated heat is transmitted back to the fuel bed to produce combustible
fuel, which then mixes with air, is heated, reacts, and liberates more heat.
Also, the combustion products when removed from the reaction zone are drawing
in fresh air thus sustaining the combustion process.

Because of the complexity of the involved heat and mass transfer phenomena,
progress in theoretical studies has been slow. As an interim approach,
much effort has therefore been devoted to laboratory studies to understand
at least some of the isolatable elements of a fire. Of these, three have
been selected here for discussion: the temperature distribution of an open
fire exposed to wind; the sometimes observed periodic contractions and
expansions of a flame; and the height of a flame.

To model these features, we consider the basic processes of a free-burning
fire: (1) Combustible fuel vapor is generated from the fuel bed by heat
supplied from the combustion zone; (2) Vapor and oxygen react in the combustion zone (visible as "flame") to liberate heat and hot combustion products;
(3) The flame and the hot, gaseous products are driven upward by buoyancy to
form a hot plume above the flame; (4) As flame and plume rise, cool air
mixes in from all sides, creating turbulent diffusion.

Clearly, inertial and bouyant forces dominate; viscous forces can be neglected.
Also, since combustion and mixing processes proceed swiftly, the conduction
of heat (a rather slow process) plays only a minor role and need not be
considered. With the additional neglection of the heat losses due to radiation, we are left only with two governing laws.

Newton's law
of inertia
$$F \triangleq \rho \ell^3 \frac{\ell}{t^2} \qquad\longrightarrow\qquad \pi_i = \frac{F}{\rho \ell^2 v^2}$$

$$t \triangleq \frac{\ell}{v}$$

Newton's law
of gravitation
expressed as
bouyant force
$$F \triangleq \Delta\rho\, \ell^3 g \qquad\longrightarrow\qquad \pi_g = \frac{F}{\Delta\rho\, \ell^3 g}$$

The ratio of these two pi-numbers leads to a special form of the densimetric
Froude number -- the bulk Richardson number (see Case Study 4)

$$Fr_d = \frac{\pi_g}{\pi_i} = \frac{\rho v^2}{\Delta\rho\, \ell g} \qquad\longrightarrow\qquad Ri = \frac{\Delta\theta\, \ell g}{\theta v^2}$$

Equv. of state (p = const), $\rho/\Delta\rho = \theta/\Delta\theta$

The last version is arrived at under the assumption that the gas density
and the density difference between hot and cold gas are directly related to
the respective gas temperatures, through the equation of state at constant
pressure.

Strict similarity, of course, requires that $\Delta\rho \triangleq \rho$ and, hence, $\Delta\theta \triangleq \theta$.
This requirement is almost impossible to achieve, however. It would entail
that $\theta/\Delta\theta = \theta'/\Delta\theta'$. Since $\Delta\theta \triangleq \theta_{hot} - \theta_{cold}$, and $\theta \triangleq \theta_{cold}$, we have the
requirement of $\theta_{hot}/\theta_{cold} = \theta'_{hot}/\theta'_{cold}$. In other words, θ'_{cold}, the ambient
air temperature of the model, is dependent on the characteristic model flame
temperature, θ'_{hot}. The model flame temperature cannot be selected independently, however, because it is the result of fuel type, fuel configuration,
ambient air temperature, etc. Hence, the ambient air temperature of the
model cannot be selected either; model tests seem impossible. A solution of
this dilemma is offered by using the same fuel for model and prototype. Then,
the temperatures θ_{hot} and θ'_{hot} should be equal. Consequently, the ambient
air temperatures, θ_{cold} and θ'_{cold}, must be kept equal, too. Under these
circumstances, the Richardson number can be simplified.

$$Ri = \frac{\Delta\theta\,\ell\,g}{\theta\,v^2}$$

(schematic diagram with conditions: $g = g'$, $\theta = \theta'$, $\Delta\theta = \Delta\theta'$, leading to $\dfrac{v}{v'} = \sqrt{\dfrac{\ell}{\ell'}}$ and $\dfrac{t}{t'} = \sqrt{\dfrac{\ell}{\ell'}}$, with $v \, \Delta \, \ell/t$)

We arrive at the simple result that the ratio of corresponding burning times will be equal to the square root of the length scale factor. The correctness of this result was checked by the following test.

7.1 Pulsating Fires and Pulsating Stars

Circular fires in quiet air often display periodic pulsations. During one cycle, the flame near its base first expands, then contracts suddenly toward its center, thereby generating an upward traveling flame bulge.

To check the validity of the basic arguments developed earlier, a series of scale model tests was performed. Ethanol fuel was burned in six geometrically similar, circular, shallow metal-pan burners, as shown schematically in Fig. 7.1. The time period of pulsation, t_p, was determined partly by a high-speed movie camera, partly by unaided counting (for $D > 30$ cm). Test data, plotted on log-log paper (Fig. 7.1), exhibit a straight-line relationship with a slope of 1:2. Hence, the model relation, $t_p/\sqrt{D} = t'_p/\sqrt{D'}$, derived from Newton's law of inertia and the buoyancy law is indeed followed by the experimental results.

The interplay between inertial and gravitational forces not only governs the pulsation of flames but also of a certain class of giant stars, called Cepheid variables. Their radii are observed to expand and contract very regularly[1] by about $\pm 10\%$,[2] with periods between a few hours and more than fifty days. Assuming that all Cepheids are homologous systems, one can deduce a simple relation between period and density.

---

[1] The period of δ Cepheids, for example, is 5 d 8 h 53 min 27.46 s, with a decrease of about 1 s in twenty years!

[2] Typically, by 2×10^6 km.

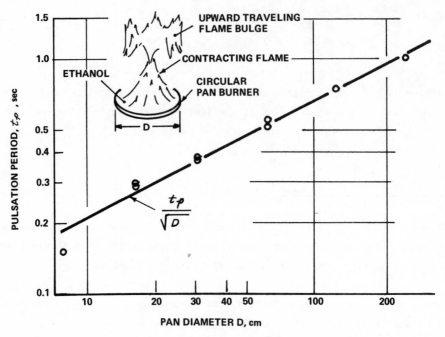

Fig. 7.1.   Test results of pulsating fire.

Newton's law of gravitation, $F = G \dfrac{m_1\, m_2}{r^2}$, can be representatively expressed as follows

$$F \triangleq G\, \frac{M m}{\ell^2}$$

where M is the total mass of the star, and m is the mass of a small element. This element is also subjected to inertial forces. In representative terms,

$$F \triangleq \frac{m \ell}{t^2}$$

These two relations lead to two principal pi-numbers.

$$\pi_g = \frac{F \ell^2}{G M m} \qquad \text{and} \qquad \pi_i = \frac{F t^2}{m \ell}$$

Both combined result in

$$\pi = \frac{\pi_i}{\pi_g} = \frac{G M t^2}{\ell^3} \xrightarrow[\substack{M \triangleq \rho \ell^3}]{\substack{G = G'}} \quad t_p^2 \rho = t_p'^2 \rho'$$

$$t = t_p$$

Hence, the product of density and period of pulsation squared is the same for all Cepheids independently of their diameters -- a relation of great value to astronomers. Figure 7.2 shows excellent agreement of the predicted relation with observed data.

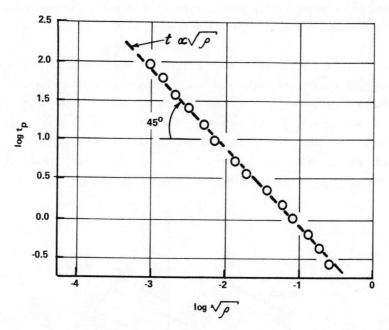

Fig. 7.2.   Empirical data on the period-density relation for Cepheid variables (after Sedov).

O AVERAGE VALUE FOR A GROUP OF CEPHEIDS

Source of Information

G.M. Byram and R.M. Nelson, "The modeling of pulsating fires," *Fire Technology* 6, 102-110 (May 1970).

L.I. Sedov, *Similarity and Dimensional Methods in Mechanics*, Academic Press, New York, 1959; Ch. V.

7.2 Fog Dispersal on Runways

The model experiment described in Case Study 7.1 proves in an indirect way the correctness of the assumption that by using the same fuel for model and prototype, the characteristic temperature of both model and prototype flames will be the same. A direct proof is delivered by the following experiment.

During W.W. II, airport runways were cleared of fog by heating the air along the runways with gasoline burners. The problem was to determine the temperature distribution downwind of the burners, as a basis for the most efficient design and arrangement of the burners. Three models of different sizes were tested in a wind tunnel (55 m long, 9 m wide, and 3.6 m high), and a field experiment was conducted at full scale. In all four tests, butane burners were arranged in a transverse line, as illustrated in Fig. 7.3.

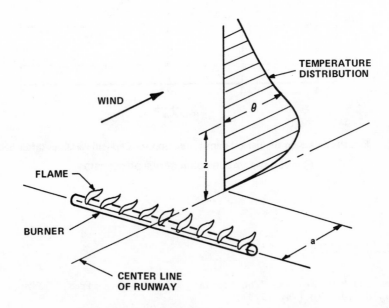

Fig. 7.3.   Experimental setup for fog dispersal.

Because losses due to heat conduction and heat radiation can be neglected, all the liberated heat of the fuel is stored in the combustion products. Then,

heat stored
in combustion $Q \triangleq c_p \rho \ell^3 \Delta\theta$ → $\Pi_c = \dfrac{Ht}{c_p \rho \ell^2 \Delta\theta}$
products

$$c_p = c_p'$$
$$\rho = \rho'$$
$$\Delta\theta = \Delta\theta'$$

same fuel

heat liberated
per unit time $H \equiv \dfrac{Q}{\ell t}$
and unit burner
length

Ri → $\dfrac{t}{t'} = \sqrt{\dfrac{\ell}{\ell'}}$

$$\dfrac{v}{v'} = \sqrt{\dfrac{\ell}{\ell'}}$$

model rule for wind speed

$$\dfrac{H}{H'} = \left(\dfrac{\ell}{\ell'}\right)^{3/2}$$

model rule for H

where H is the heat liberation per unit time and unit length of the burner line. Using the same fuel for model and prototype leads again to the assumption of equal representative temperatures of model and prototype plumes and, hence, of equal densities and specific heats. As a result, two model design rules ensue, one for H , and one for wind speed, v.

In all four tests, the fuel supply of the butane burners was adjusted so as to generate the required amount of heat, H . To achieve the required wind speed, v, the wind speed of the field test was measured and the tunnel speeds were set at corresponding levels. Table 7.1 lists H and v together with geometrical data.

TABLE 7.1. Input Data for Model Experiments.

LENGTH SCALE FACTOR	WIND SPEED v m/s	LIBERATED HEAT H J/(m s)[1]	DISTANCE a m
1	3.8	72.0×10^4	68.0
15	1.0	1.24×10^4	4.5
30	0.7	0.44×10^4	2.25
60	0.5	0.15×10^4	1.14

The vertical temperature distribution at centerline was measured at distance "a" downstream. As shown in Fig. 7.4, the temperature rise was indeed the same at all corresponding points of the three models and the prototype.

Source of Information

A.O. Rankine, "Experimental studies in thermal convection," *Proc. of the Phys. Soc.* 63, 364B, 225-251 (Apr 1950).

The same work is also briefly described in: G.I. Taylor, "Fire under the influence of natural convection," *The Use of Models in Fire Research*, W.G. Berl (ed), Pub. No. 786, Natl. Acad. of Sci., Natl. Res. Council, Washington, D.C., 1961.

---

[1]The original data in the Source of Information were given in Therms. Here, one Therm was taken to be 10^5 British thermal units.

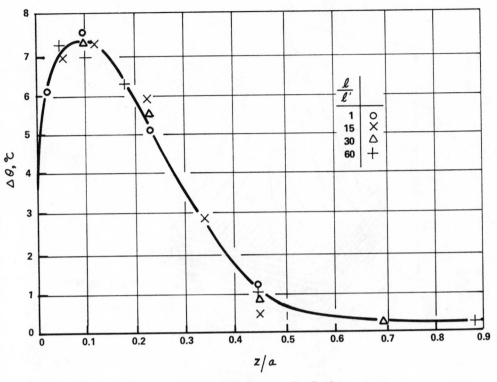

Fig. 7.4.   **Vertical temperature distribution .**

7.3   Flame Height

The developed model design rules were applied to the rate of burned fuel and
the associated flame height, both measured at constant fuel supply. Model
and prototype consisted again of the same fuel, here sticks of spruce
stacked in square cribs of length D , Fig. 7.5. By varying the number of
sticks, the rate of fuel supply, $\dot{m}$, was varied. The rate of burning was
recorded by constantly weighing the crib.

After a short transition period, the burning rate remained constant for a
time, during which combustion was mainly maintained by the volatiles of the
wood. Later, when carbon started to burn, the rate of burning fell, and the
test was stopped.

The height of the flame, L , was measured photographically.

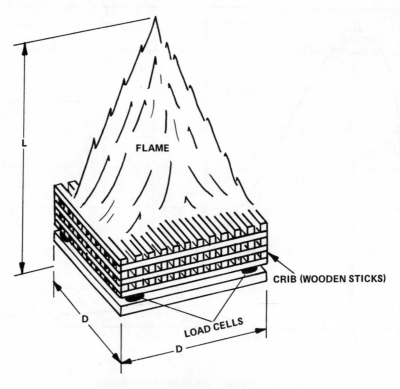

Fig. 7.5.   Experimental fire.

The previously derived model design rule (developed from the Richardson number) can be easily adapted to include the rate of mass loss, $\dot{m}$. Actually, the model design rule provides only for the gaseous fuel produced from the wood, but during the period of constant burning rate (the period of interest here), the mass of these gases is proportional to the mass loss of the crib. Therefore,

$$\underline{Ri} \longrightarrow \frac{v}{v'} = \sqrt{\frac{\ell}{\ell'}} \longrightarrow \frac{\dot{m}^2}{\ell^5} = \frac{\dot{m}'^2}{\ell'^5}$$

$$\text{model rule for } \dot{m}$$

$$\text{crib mass loss } \dot{m} \triangleq \rho \frac{\ell^3}{t} \longrightarrow \frac{v}{v'} = \frac{\dot{m}\,\ell'^2}{\dot{m}'\,\ell^2}$$

$$t \triangleq \frac{\ell}{v}$$

$$\rho = \rho'$$

$$\text{same fuel}$$

The non-dimensional flame height, L/D, would become a function of the developed design rule if all underlying assumptions are correct. As seen from Fig. 7.6, excellent agreement among data from tests of various sizes was obtained; thus the basic assumptions appear to be justified.

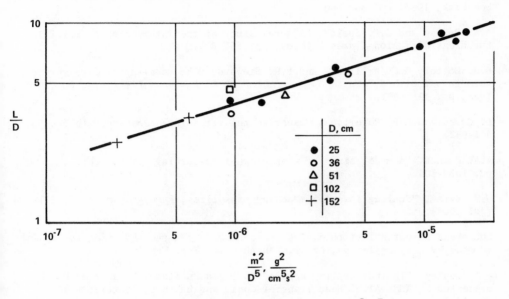

Fig. 7.6.   Dimensionless flame height versus $\dot{m}^2/D^5$.

Source of Information

P.H. Thomas, C.T. Webster, and M.M. Raftery, "Some experiments on buoyant diffusion flames," *Combustion and Flame* 5, 4, 359-367 (Dec 1961).

Selected Related References

W.G. Berl (ed), *Intl. Symp. on the Use of Models in Fire Research*, Publ. 786, Natl. Acad. Sci., Nat. Res. Council, Washington, D.C., 1961.

W.L. Fons, H.B. Clements, and P.M. George, "Scale effects on propagation rate of laboratory crib fires," *Ninth Symp. (Intl.) on Combustion*, Academic Press, New York, 1963, pp. 860-866.

A.A. Putnam and C.F. Speich, "A model study of the interaction of multiple turbulent diffusion flames," ibid., pp. 867-877.

H.C. Hottel, G.C. Williams, and F.R. Steward, "The modeling of fire spread through a fuel bed," *Tenth Symp. (Intl.) on Combustion*, Pittsburg, Pa., 1965, pp. 997-1007.

D. Gross and A.F. Robertson, "Experimental fires in enclosures," ibid., pp. 931-942.

A.A. Putnam, "A model study of wind-blown free-burning fires," ibid., pp. 1039-1046.

G.M. Byram, "Scaling laws for modeling mass fires,"*Pyrodynamics* 4, 3, 271-284 (Jul 1966).

Southwest Research Institute, *The Use of Models for the Investigation of Fire Spread*, by L.A. Eggleston, et al, San Antonio, Tex, Feb 1966.

J.A. Copley, "Thermal scaling applied to luminous flames," Paper 67-HT-60, presented at ASME-AIChE Heat Transfer Conf. and Exhibit, Seattle, Wash, Aug 1967.

F.A. Williams, "Three unconventional approaches to experimental scaling of mass fires," *Proc. Tripartite Technical Cooperation Program, Panel N-3 (Thermal Radiation), Mass Fire Research Symp.*, Oct 1967, Defense Atomic Support Agency, DASA-1949, Santa Barbara, Calif, pp. 106-126.

U.S. Naval Radiological Defense Lab., *An Experimental Test of Mass Fire Scaling Principles*, by W.J. Parker, R.C. Corlett, and B.T. Lee, NRDL TR-68-117, San Francisco, Calif, Jul 1968.

U.S. Naval Radiological Defense Lab., *Mass Fire Scaling with Small Electrically Heated Models*, by B.T. Lee, NRDL TR-69, San Francisco, Calif, May 1969.

U.S. Naval Radiological Defense Lab., *Camp Park Mass Fires*, by C.P. Butler, US NRDL TR-69-A, San Francisco, Calif, Aug 1969.

Stanford Res. Inst., *Laboratory Scaling of the Fluid Mechanical Aspects of Mass Fires*, by B.T. Lee, SRI Proj. PYU-8150, Menlo Park, Calif, Aug 1970.

W.D. Weatherford, "Scaling of flames above free-burning structural models," *Combustion and Flames* 14, 1, 21-29 (Feb 1970).

Ministry of Technology, Safety in Mines Res. Establishm. (SMRE), *Fires in the Timber Lining of Mine Roadways: a Comparison of Data from Reduced-Scale and Large-Scale Experiments*, by A.F. Roberts, PB 197 096, Buxton, England, 1970.

R. Friedman, "Aerothermodynamics and modeling techniques for prediction of plastic burning rates," *J. Fire and Flammability* 2, 240-256 (Jul 1971).

Stanford Res. Inst., *Modeling the Dynamic Behavior of Building Fires*, by B.T. Lee, Menlo Park, Calif, Aug 1971.

S.L. Lee, "Fire research," *Appl. Mech. Rev.* 25, 5, 503-509 (May 1972).

T.E. Waterman, "Room flashover - model studies," *Fire Technology* 8, 4, 316-325 (Nov 1972).

G. Heskestad, "Modeling of enclosure fires," *Fourteenth Symp. (Intl.) on Combustion*, Combustion Inst., Pittsburg, Pa., 1973, pp. 1021-1030.

H.W. Emmons, "Heat transfer in fire," *Trans. ASME, J. Heat Transfer* 95, C2, 145-151 (May 1973).

For further references on modeling fires and flames, see *Fire Research Abstracts and Reviews*, Natl. Academy of Science, 1958 --.

CASE STUDY 8

Architectural Acoustics

Although the acoustic quality of a concert hall is ultimately judged by the audience, a number of quantitative criteria are also available. One of them is the reverberation time, T, defined as the time required for the rms sound pressure to decay 60 dB after the source is stopped.[1] The desired reverberation time for concert halls is about 1.8 seconds for frequencies between 500 and 1000 Hz, and 1.3 seconds for higher frequencies; but predicting the reverberation time theoretically is difficult, for it depends on the acoustical properties and the geometrical shapes of the walls, the ceiling, the floor, the stage, and the seats. Variation of these design parameters greatly changes T and its distribution over frequency, to affect such acoustical criteria as warmth, liveness, and brilliance.

There are other, more qualitative criteria that completely resist analytical treatment, such as spatial uniformity of sound, intimacy, loudness, balance, freedom of echoes, diffusivity, and definition. Therefore, model experiments that would permit a parametric study of all acoustical criteria including the subjective appraisal of musical renditions are highly desirable and have indeed been developed into an effective design tool for concert halls.

---

[1] The energy rate of a sound wave passing through a unit area is called acoustic intensity I. It is conventionally compared with the minimum acoustic intensity audible to the average human ear (at 10^3 Hz), $I_o = 10^{-16}$ W/cm^2. Since the ratio I/I_o can vary from 1 to 10^{14} in audible range, a logarithmic scale is used in form of the sound intensity level, IL = 10 log (I/I_o), expressed in decibel (dB). The decibel notation is also used for the sound pressure level, SPL, of a sound wave. Here, the rms sound pressure equivalent to I_o is $p_o = 20 \mu$N/m^2. With $I/I_o = (p/p_o)^2$, SPL = 20 log (p/p_o) = IL, in dB.

8a. Modeling Sound Waves

Physically, sound can be described in terms of longitudinal waves accelerating
air molecules periodically in fore and aft direction, thereby producing alter-
nate regions of compression and rarefaction. Hence, sound is governed by
inertial and compression energies. Viscous loss energies can be neglected,
and so can heat losses since very little heat flows away from a compressed
part before it is rarefied. Therefore, the law of adiabatic compression can
be applied. Under these circumstances, two pi-numbers evolve.

Inertial force
of air (Newton's
law of inertia)

Adiabatic air
compression

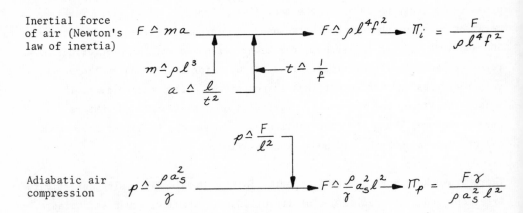

a_s is the speed of sound; f is the frequency.

Elimination of the representative force delivers

$$\pi_1 = \frac{\pi_p}{\pi_i} = \frac{\ell^2 f^2 \gamma}{a_s^2}$$

$a_s = a_s'$

$\gamma = \gamma'$

same air (same temp.)

$$\frac{f}{f'} = \frac{\ell'}{\ell}$$

frequency
model rule

Consequently, if the same air is used for model and prototype, the frequencies
of the model sound must be ℓ/ℓ'-times higher than the corresponding prototype
frequencies.

8b. Modeling Sound-Absorbing Materials

A tone produced in a closed room usually decays quickly, mainly through
absorption within the walls and the objects in the room (including the
audience). Wall absorption is characterized by the (dimensionless) absorption coefficient, α , defined as the ratio of absorbed and incident sound
energy at a given frequency.

The coefficient of absorption can be measured in various ways. In the
Source of Information, the reverberation room method was used. The test
room had a volume of 255 m^3 and very hard walls, hence a long reverberation
time, T_0 , in the empty state. The sample material was arranged along the
walls; and the reverberation time, T_1 , of test sounds (emitted by loud-
speakers) was measured at various frequencies. The coefficient of absorption of the test material was then computed from T_0 and T_1 by a theoretical
formula.[1] Since this formula is not exact, only approximate values can be
obtained. Occasionally, α -values larger than unity are computed, see
Fig. 8.2 and 8.4. For mere comparisons of different materials, these
inaccuracies are not important, however.

The three common types of absorbing materials are porous structures, thin
panels or membranes, and cavity resonators. Porous structures absorb sound
most effectively at higher frequencies, whereas panels and resonators produce
high absorption around their natural frequencies, usually below 1000 Hz. By
combining porous structures with panels or resonators, effective sound absorption at almost any desired frequency can be achieved.

To model the interaction of sound waves with absorbing wall material, the
kinetic and the pressure energy of the sound waves must be modeled, as discussed earlier. If the wall structure is flexible (e.g., panels) and, hence,
able to vibrate, the elastic and inertial energies of the structure come into
play, too.

---

[1]Sabine-Eyring formula.

Ultimately, the absorbed energy will be dissipated as heat, owing to viscosity, heat conduction, and internal damping of the vibrating structure -- phenomena that are difficult to model because their governing laws are either not well known or, if known, lead to conflicting design requirements. Take, as an example, the pi-number expressing the modeling requirements for viscous losses. Viscous losses are governed by the law of shear stresses of a Newtonian fluid.

Shear stress of Newtonian fluid

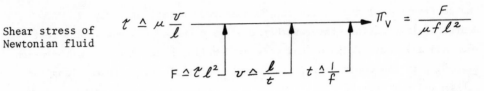

This pi-number, combined with Π_i , results in

$$\Pi_2 = \frac{\Pi_V}{\Pi_i} = \frac{\rho \ell^2 f}{\mu} \xrightarrow[\substack{\rho = \rho' \\ \mu = \mu'}]{\substack{\text{same air} \\ \text{(same temp.)}}} \frac{f}{f'} = \left(\frac{\ell'}{\ell}\right)^2$$

This result is in conflict with the requirement of $f/f' = \ell'/\ell$ derived earlier from the energies governing a sound wave. Hence, faithful modeling of viscous losses *and* sound waves appears to be impossible, and relaxations are in order.

<u>Porous material</u>. When a sound wave strikes porous material, the sound pressure will excite the air contained in the fine canals of the material. To overcome the frictional resistance between vibrating air and solid material, energy is taken away from the sound wave.

To model the viscous losses in all the individual capillaries and pores is impossible, not only because the pi-number Π_2 conflicts with Π_1 , but also because the variations of the pores in shape, size, direction, their interconnections, and their distribution cannot be scaled down in detail. Instead, a statistical view must be taken that would permit modeling viscous losses without conflicting with Π_2 and without requiring faithful scaling of the material's structure.

Pores can be regarded as parallel capillaries with a representative radius, R, and a representative length, ℓ (Fig. 8.1). The length ℓ also represents all external dimensions of the material; for instance, the thickness. The representative radius is determined indirectly by a d-c permeability test in which the permeability, k (the measure of the ease with which a gas flows through a porous medium), can be pictured as the representative pore area.[1] Thus, in representative terms, $k \triangleq R^2$. Then, the pressure force can be expressed as

$$F \triangleq p R^2 \triangleq p k$$

and the viscous force as
$$F \triangleq \mu R \ell \frac{v}{R} = \mu \ell v$$

Elimination of the representative force results in the pi-number

$$\pi_3 = \frac{p k}{\mu \ell v}$$

Now, the pi-number π_p calls for $p = p'$ (same air for model and prototype); and the pi-number π_1, for $\ell \triangleq \frac{1}{f}$, which is equivalent to $v = v'$. Consequently, π_3 results into the statement of $k/\ell = k'/\ell'$. The permeability is usually determined by a flow test.[1]

$$k = \frac{\dot{Q} \mu}{A \, dp/dL}$$

Since $\dot{Q} = v A$, where v is the flow speed, it follows that $k/\ell \propto v/p$. The ratio p/v is called the d-c acoustical resistance; it is usually expressed in Rayls (equivalent to Ns/m^3). For many materials, the a-c and d-c acoustical resistances are practically the same. However, if the sound energy is absorbed not only by viscous damping but also by internal damping of the material, then the a-c resistance may be much larger than the d-c resistance. For these materials, our considerations are not valid.

---

[1] See Section "Restriction to Gross Effects -- Spatially Integrated Effects," Part I.

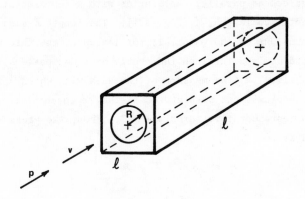

Fig. 8.1.   Representation of porous materials.

Acousticians often claim that to characterize sound absorption, the d-c
resistance must be supplemented by another measurable quantity, the porosity,
ϕ, or the volume of pore space divided by the total volume of the medium.
In representative terms, $\phi \triangleq R^2/\ell^2$. The requirement of $\phi = \phi'$, however,
conflicts with the requirement of $\ell/k = \ell'/k'$ because k represents R^2,
as introduced earlier. Consequently, the requirement of equal porosity must
be relaxed.

The porous prototype material used (Source of Information) was a felt material
with the brand name of Sillan-Filz SF 57. The d-c acoustical resistance was
50 Rayls, the porosity 0.80. Its absorption coefficient was determined at
various frequencies by the reverberation room method. Figure 8.2 shows α
as a function of fd, where d = 5.0 cm, the material thickness.

As model material, a heavy felt used for saddle blankets was tested. Since
a length scale factor of 10 was selected, the model felt was made 0.5 cm
thick. Its acoustical resistance, 63 Rayls, was considered adequately close
to the prototype's resistance (and so was its porosity of 0.85).

The absorption coefficient of the model material should equal that of the
prototype at equal fd-values. To measure the absorption coefficient of the
model felt under conditions similar to those employed for the prototype, a
small reverberation room was built with lengths ten times smaller than the

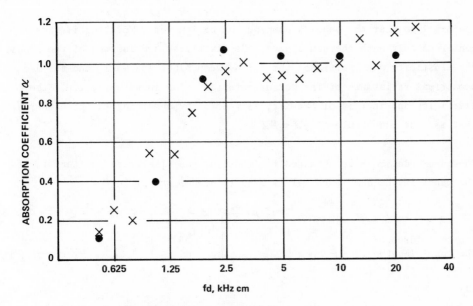

SILLAN FILZ SF57, THICKNESS d = 5 cm, d-c ACOUSTICAL
RESISTANCE 50 Rayls, POROSITY 0.80

HEAVY FELT, THICKNESS d = 0.5 cm, d-c ACOUSTICAL
RESISTANCE 63 Rayls, POROSITY 0.85

Fig. 8.2.   Sound absorption coefficient α of porous material.

corresponding lengths of the prototype reverberation room. The air of the
model room was dried down to 3% humidity, for reasons explained later. The
measured coefficient plotted versus fd values of the model agrees remarkably
well with the prototype values, Fig. 8.2.

Thin Panels. Thin panels mounted at some distance from a solid wall convert
sound energy into vibrational energy, with the air between panel and wall
acting as spring. The greatest movement and, hence, the greatest absorption
is obtained at the resonance frequency of the panel-air system. If the panel
is thin enough to act as a membrane, its stiffness can be neglected; if not,
its stiffness must be taken into account together with its mass.

Ultimately, the absorbed energy is dissipated as heat by virtue of the
system's internal damping, which is usually rather small. To increase internal
damping, the space between panel and wall is partly or completely filled with
soft porous material, whose internal resistance is large compared to that of

the panel so that the panel's damping can be ignored. Modeling is then concerned only with the sound wave, the elasticity and inertia of the panel, the elasticity and inertia of the air between panel and wall, and the d-c acoustical resistance of the porous material. The sound wave, i.e., the elasticity and inertia of the air, is modeled by observing the pi-number π_1 ; that is, the model rule of $f\ell = f'\ell'$.

The same model rule is obtained for modeling the elasticity and inertia of the panel if the same material is used for model and prototype:

$$m \triangleq \rho \ell^3 \qquad a \triangleq \frac{\ell}{t^2} \qquad t \triangleq \frac{1}{f}$$

Inertial force of the panel (Newton's law of inertia)
$$F \triangleq ma \longrightarrow \pi_i = \frac{F}{\rho \ell^4 f^2}$$

Elastic force of the panel (Hooke's law)
$$F \triangleq E \ell^2 \epsilon \longrightarrow \pi_e = \frac{F}{E \ell^2 \epsilon}$$

$$\pi_4 = \frac{\pi_i}{\pi_e} = \frac{E \epsilon}{\rho \ell^2 f^2} \xrightarrow[\text{geometrical similarity}]{\epsilon = \epsilon'} \xleftarrow[\text{same material}]{E = E' \\ \rho = \rho'} \frac{f}{f'} = \frac{\ell'}{\ell}$$

Figure 8.3 shows model and prototype results of panel tests. The prototype panel consisted of plywood, 1.5 cm thick, with a density of 0.5 g/cm^3. It was mounted 5 cm from the wall, with the air space partly filled with Sillan mat. The model panel was made from veneer glued to construction paper, 1.5 mm thick with a density of 0.5 g/cm^3. It was mounted 5 mm from the wall; the air space was partly filled with polyester foam of the same d-c acoustical resistance as the prototype material.

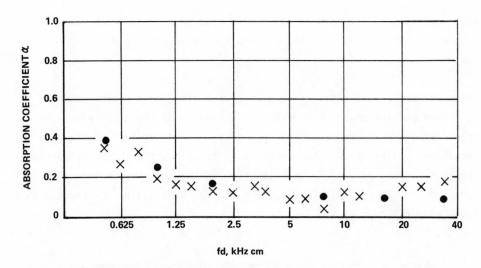

Fig. 8.3.   Sound absorption coefficient α of thin panels.

The coefficient of sound absorption was measured as a function of frequency: of the prototype, in the large reverberation chamber; of the model panel, in the model chamber, as described earlier. The agreement between model and prototype was quite good.

Cavity Resonators. A cavity resonator acts as a spring-mass system: the pressure of the sound waves striking the wall vibrates the mass of the air around and in the cavity's throat, with the air in the cavity acting as a spring. At resonance frequency, the movements are quite strong; their energy is dissipated mainly by porous material inside the cavity and also by viscous friction within the throat. If we neglect the throat friction and model only the inertia and the elasticity of the air and the absorption of the porous material, then only π_1 and π_3 apply, as discussed before.

Prototype tests in the large reverberation chamber were performed with a perforated plaster-of-Paris plate mounted 5 cm from the wall. The holes, 1 cm in diameter and covering 14.8% of the surface, acted as throats of so

many "cavities," with each cavity represented by a certain air space behind
a hole. The plate was 1 cm thick, with the space between plate and wall
filled with rock wool.

The model was a perforated steel plate, with all linear dimensions made
ten times smaller than the corresponding lengths of the prototype. The
space between plate and wall was filled with polyester foam of the same d-c
acoustical resistance as the prototype material. Tests in the model reverberation chamber were performed with frequencies ten times higher than the
prototype's. Test results are plotted in Fig. 8.4; again, agreement between
model and prototype is good.

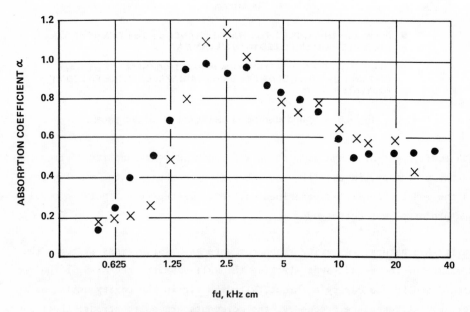

fd, kHz cm

● PLASTER OF PARIS PLATE; HOLE DIAMETER 1 cm; PERCENT OF AREA
COVERED WITH HOLES 14.8%; PLATE THICKNESS 1 cm; WALL DISTANCE
d = 5 cm, AIR SPACE FILLED WITH MINERAL WOOL

✕ STEEL PLATE; HOLE DIAMETER 0.1 cm; PERCENT OF AREA COVERED
WITH HOLES 15%; PLATE THICKNESS 0.1 cm; WALL DISTANCE d = 0.5 cm,
AIR SPACE FILLED WITH POLYESTER FOAM

Fig. 8.4.   Sound absorption coefficient α of perforated plates.

8c. Modeling Air Attenuation

Although wall absorption is the most important source of energy dissipation, it is not the only one. A small amount of sound energy penetrates the walls and escapes; however, the amount will be negligibly small if the walls are thick and heavy. Other negligibly small amounts are dissipated through viscous air friction at the walls and internal air friction. Yet another source of absorption is the excitation of the air's oxygen molecules to vibrations in the presence of small amounts of water vapor. This effect can be pronounced. Even in a room with perfectly reflecting walls, a tone would still die out quickly at relative humidities larger than a few per cent.

Air attenuation is expressed by the attenuation coefficient m, which describes the attenuation of sound intensity per unit distance.

$$I = I_o e^{-m x}$$

where I is the sound intensity at distance x, and I_o is the sound intensity at $x = 0$. Conversion of this relation results in

$$m = -\ln(I/I_o)/x \text{, neper/meter}$$

The neper is a dimensionless unit; one neper corresponds to a reduction to $1/e$ of the reference value.

The attenuation coefficient depends on frequency and air humidity, as indicated in Fig. 8.5.

The reverberation time, T_a, based on air attenuation alone is computed from the attenuation relation by replacing the distance x by the product of $a_s T_a$. Then,

$$I/I_o = e^{-m a_s T_a}$$

The reverberation time is defined for $\log(I/I_o) = -6$, so that

$$m \, a_s T_a = 6 \ln 10; \text{ and with } a_s = 331 \text{ meter/sec, } T_a = 0.04/m \text{, in sec.}$$

If T_a is much larger than the total desired reverberation time of about 2 seconds, the influence of air attenuation can be ignored. Consequently,

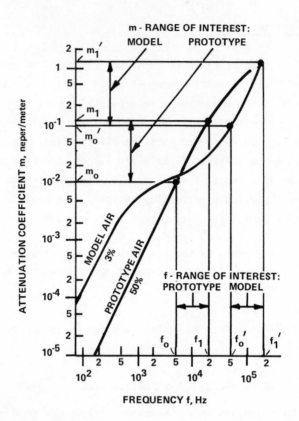

Fig. 8.5.   Air attenuation coefficient m at relative air humidities of 3% and 50% .

m-coefficients smaller than, say, m_o = 0.04/(4 sec) = 10^{-2} neper/meter are
of no interest. Figure 8.5 indicates that under these circumstances, air
attenuation at a relative humidity of 50% is of importance only for
frequencies beyond f_o = 5 kHz, up to f_1 = 20 kHz, the limit of audible
sound. The m-coefficient associated with f_1 is, according to Fig. 8.5,
m_1 = 0.15 neper/meter.

For a length scale factor of 10, the corresponding model frequency range
lies between f_o' = 50 kHz and f_1' = 200 kHz. Since the attenuation coefficient
is inversely proportional to distance, the model m-range lies between
m_o' = 0.1 neper/meter and m_1' = 1.5 neper/meter. Figure 8.5 shows that the
thus specified model ranges of f' and m' coincide with the curve for 3%
relative humidity. Therefore, the model air was dried down to about 3%.

8d. Modeling a Concert Hall

After completion of the Meistersingerhalle in Nuremberg, Germany, various
acoustical tests were performed and their results compared with those obtained from model tests. The length scale factor was 10:1. Great care was
exercised in modeling the sound absorption of walls, seats, audience, etc.
(the audience was simulated by egg cartons with an α per person of about 0.4).
The model air was dried to 4% relative humidity, which was considered close
enough to the desired 3%. To generate non-directional sounds of a wide frequency spectrum, a spark source was used for both model and prototype. The
property of non-directivity of the sound source is important for determining
the reverberation time, for most natural sound sources are reasonably non-
directive, particularly when organized in groups. The sound was picked up by
omni-directional microphones, and its decay was recorded and analyzed. The
analysis yielded reverberation time as a function of frequency between 0.1
and 100 kHz, for the model as well as for the prototype. The agreement of
reverberation times at corresponding frequencies was very good, as illustrated in Fig. 8.6.

Many other, more subjective tests were performed. Of them, we mention only
the so-called room perception ("Hoersamkeit") tests. Short excerpts of a
string quartet and a speech were recorded in an "echo-free" room and played
back (the music, stereophonically); in the model, at ten times higher speed;
and in the Meistersingerhalle and five other halls, at normal speed. The
played-back pieces were then recorded at six different seat locations in the
model and the concert halls. These recordings were presented in pairs to
ten critical listeners, a pair consisting of a model recording and one of the
concert halls recordings. The task of the listeners was to match the model
recording and the original Meistersingerhalle recording. Figure 8.7 gives
the results.

On the average, eight out of ten listeners identified the Meistersingerhalle
as the hall whose acoustical signature was closest to that of the model -- a
very convincing demonstration of the predictive power of acoustical model
tests.

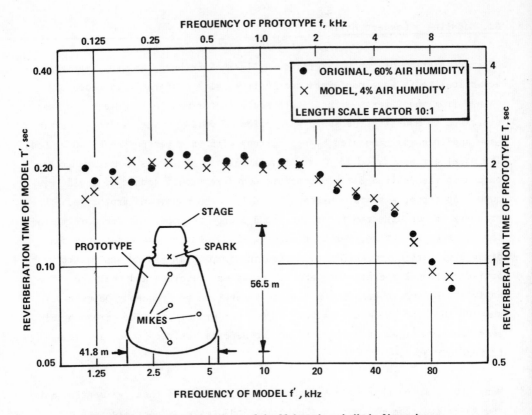

Fig. 8.6.   Reverberation times of the Meistersingerhalle in Nuremberg.

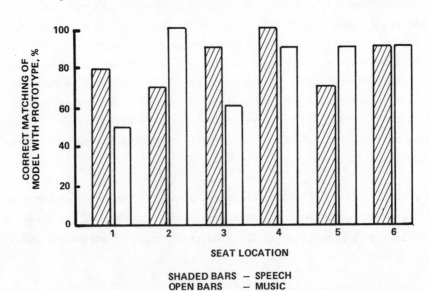

Fig. 8.7.   Subjective matching of model and prototype recordings.

Source of Information

D. Brebeck, R. Bücklein, E. Krauth and F. Spandöck, "Akustisch aehnliche Modelle als Hilfsmittel fuer die Raumakustik" (Acoustically similar models as an aid in architectural acoustics), *Acustica* 18, 4, 213-226 (1967).

Related References

H. Winkler, "Acoustical measurements on model and prototype of the Theatre 'Kosmos' in Berlin"(in German), *Acustica* 14, 6, 347-355 (1964).

Proc. Fifth Intl. Congr. Acoust., Imprimerie Georges Thone, Liege, France, 1965.

V.L. Jordan, "Acoustical considerations concerning New York State Theatre," *J. Audio Eng. Soc.* 13, 2, 98-103 (Apr 1965).

P.S. Veneklasen, et al, "Acoustical modeling in architectural acoustics," *J. Acoust. Soc. Am.* 37, 6, 1188 (Jun 1965) (abstract of paper presented at 69th Meeting of Acoust. Soc. Am.).

B.G. Watters, et al., "Role of model testing in the acoustical design of auditoria," *J. Acoust. Soc. Am.* 40, 5, 1245 (Nov 1966) (abstract of paper presented at 72nd Meeting of the Acoust. Soc. Am.).

J.H. Janssen, "Model experiments on sound transmission from engine room to accommodation in motorships," *Intl. Shipbuilding Progr.* 14, 159, 414-420 (Nov 1967).

E. Meyer, H. Kuttruff, and N. Roy, "Model investigations of the acoustical field of the city hall in Göttingen"(in German), *Acustica* 14, 3, 132-142 (1967/68).

W. Gabler, et al., "Investigations on an acoustically 'ideal' room" (in German), *Acustica* 19, 5, 264-270 (1967/68).

W. Kuhl, "Comments on the paper 'Investigations on an acoustically ideal room' by W. Gabler, et al.," (in German), *Acustica* 20,3, 184-186 (1968).

B.F. Day, "A tenth scale model audience," *Appl. Acoustics* 1, 2, 121-135 (Apr 1968).

W. Reichard, P. Budach, and H. Winkler, "Room acoustical model investigations with the pulse test method for the rebuilt Congress and Concert Hall in the 'Haus des Lehrers' at Berlin" (in German), *Acustica* 20, 3, 149-158 (1968).

P.S. Veneklasen and J.P. Christoff, "Acoustical modeling of auditoriums: correlation with full-scale Saratoga Pavilion," *J. Acoust. Soc. Am.* 46, 1, 98 (1969) (abstract of paper presented at 77th Meeting of Acoust. Soc. Am.).

L.L. Beranek, "Acoustical modeling as a tool in problem solving," *J. Audio Eng. Soc.* 17, 2, 151-155 (Apr 1969).

B. Day and R.J. White, "A study of the acoustic field in landscaped offices with the aid of a tenth-scale model," *Appl. Acoustics* 2, 3, 161-183 (Jul 1969).

V.O. Knudsen, "Model testing of auditoriums;" V.L. Jordan, "Acoustical criteria for auditoriums and their relations to model techniques;" P.S. Veneklasen, "Model techniques in architectural acoustics;" *J. Acoust. Soc. Am.* 47, 2, Pt. 1, 401-412, 419-423 (Feb 1970).

U.S. Office of Naval Research, *Model Studies in Acoustics*, by R.O. Rowlands, C-31-70, Branch Office, London, Engl., Dec 7, 1970.

G. Koopman and H. Pollard, "Model studies of Helmholtz resonances in rooms with windows and doorways," *J. Sound and Vibration* 16, 4, 489-504 (1971).

L.W. Hegvold, "A 1:8 scale model auditor," *Appl. Acoustics* 4, 4, 237-256 (Oct 1971).

Proc. 7th Intl. Congr. Acoust., Akademiai Kiado, Budapest, 1971.

P.S. Veneklasen, "Acoustical modeling -- progress report," *J. Acoust. Soc. Am.* 51, 1, Pt. 1, 128 (1972) (abstract of paper presented at 82nd Meeting of Acoust. Soc. Am.).

W. Schmidt, "Room acoustical planning with the help of models" (in German), *Wiss. Zeitschr. Univ. Dresden* 22, 5, 803-809 (1973).

H.D. Harwood and A.N. Burd, "Acoustic scaling of studies and concert halls," *Acoustica* 28, 6, 330-340 (1973).

V.L. Jordan, "Auditoria acoustics," *Appl. Acoustics* 8, 3, 217-235 (Jul 1975).

Modeling Sound Absorption and Insulation

N.F. Egorov, "Scale-modeling of sound-absorbing layers of fibrous materials," *Soviet Physics-Acoustics* 13, 3, 320-323 (Jan-Mar 1968) (on page 322, the thickness of the layers was reported to range from 0.5 to 10 mm, undoubtedly a misprint; for to be consistent with the figures, the dimensions should have been 0.5 to 10 cm.).

F.H. van Tol, "A model study of damping layers applied in ship structures," *Proc. Fifth Intl. Congr. on Acoustics* 1965, Liege, D.E. Commins (ed), Vol. 1a, Paper F63, Imprimerie Georges Thone, Liege, 1965.

S. Zurnatzis, "Investigations of sound insulation problems by models," (in German), ibid, Paper F67.

D. Brebeck, "Simulation of acoustically similar wall materials for acoustical model tests in rooms" (in German), ibid., Paper H63.

E. Meyer, H. Kuttruff, and W. Lauterhorn, "Model tests for determination of the degree of sound absorption in reverberation rooms," (in German), *Acustica* 18, 1, 21-32 (1967).

R.H. Tomren and P. Brandtzaeg, "Noise-reduction study in a scaled model acoustic facility," *J. Acoust. Soc. Am.* 44, 1, 359 (Jul 1968) (abstract of paper presented at 75th Meeting Acoust. Soc. Am.).

C.L.S. Gilford, "Sound insulation studies using model techniques," *Model Studies in Acoustics*, U.S. Office of Naval Res., Branch Office, London, England, C-31-70 (abstract of paper presented at one-day conference organized by British Acoust. Soc., London, Nov 12, 1970).

R.K. MacKenzie, "Fifth-scale model studies of sound insulation using simulated building elements," ibid.

A. Behar, "Measurement of transmission loss of small dimension samples," *Acustica* 25, 5, 326-330 (1971).

H. Tachibana and K. Ishii, "Simulation of the sound absorption characteristics of the absorbents for acoustic scale model experiments," (Japanese; English abstract), *J. Acoust. Soc. Jap.* 28, 4, 169-175 (1972).

A. Cops, H. Myncke, and E. Lambert, "Sound insulation of glass by means of scale models," *Acustica* 31, 3, 142-149 (Sep 1974).

CASE STUDY 9

Induction Furnace

The motions of molten metal in induction furnaces are often utilized to support desired physical and chemical changes in the charge. To take full advantage of the flow, however, the speed and trajectories of the molten metal must be known. Theoretical prediction of the flow pattern is nearly impossible because of the complex ways the many controlling parameters interact. On the other hand, direct measurements in full-size furnaces are very difficult because of the high temperatures and the possible aggressiveness of the melt. Scale model tests are therefore attractive.

The following scale model tests were set up to determine the relative importance of all participating laws, so that model rules could be established to reflect only those laws that are the most important. The tests planned for this purpose were self-modeling tests, i.e., tests in which the length scale factor and the scale factors of material properties were unity. Thus, the exploration of the complex phenomenon was reduced to varying quantities other than size and materials, such as times, temperatures, and electric currents. This appeared to be an appropriate technique for the identification of the weaker laws -- the intent of the following study. With the weak laws eliminated, more extensive model tests could then be designed that would permit variation of all parameters including linear dimensions and materials.

The investigations were carried out on a model of an ordinary induction furnace consisting of a crucible with an inner diameter of 17 cm, a single phase inductor with a uniform current distribution, and an aluminum charge, as indicated in Fig. 9.1. All measurements were made after the aluminum was melted and superheated to a constant temperature. The metal was then in rapid motion, with its free surface assuming the shape of a meniscus.

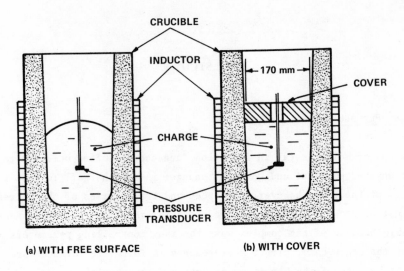

Fig. 9.1. Model induction furnace (schematic).

Since the steady-state process is maintained by the electro-magnetic field of
the inductor and influenced by the mechanical and thermal energies of the melt,
three groups of laws can be assumed to interact: electrodynamics, fluid
dynamics, and heat transfer.

Electrodynamic Laws. Both electric and magnetic fields are present to exert
forces on the moving charges in the molten metal. The expression for the
total force is known as the Lorentz force equation

$$\vec{F} = q\,(\vec{E} + \vec{v} \times \vec{B})$$

The first term constitutes the force induced by the electric field. Expressed in representative terms, it assumes the form of $F \triangleq q E$. To replace
the electric field strength, E , by the measurable quantity of electric current, i , Ohm's law is introduced, $\vec{j} = \sigma \vec{E}$, or, in representative terms,
$j \triangleq \sigma E$. Then,

$$F \triangleq q E \longrightarrow \Pi_{el} = \frac{F \ell^2 \sigma}{q\,i}$$

Ohm's law, $E \triangleq \dfrac{j}{\sigma}$

Definition, $j \triangleq \dfrac{i}{\ell^2}$

The second term constitutes the force induced by the magnetic field. Its representative form is $F \triangleq g v B$. This expression can be modified with the help of Maxwell's equation of magnetic induction, $\vec{\nabla} \times \vec{H} = \vec{j}$, and the relation between B and H , $\vec{B} = \mu \vec{H}$. Transformed into relations among representative quantities, these two laws assume the forms of $H/\ell \triangleq j$ and $B \triangleq \mu H$. Then,

$$F \triangleq g v B \qquad \qquad \qquad \qquad \qquad \pi_{ma} = \frac{F}{g f \mu i}$$

$$B \triangleq \mu H \qquad \qquad j \triangleq i/\ell^2 \qquad v \triangleq \ell f$$

$$\text{Maxwell's law, } H \triangleq \ell j$$

The pi-number π_{ma} can be further modified with the help of the representative relation of $g \triangleq it$, derived in Note 1 of the Appendix at the end of Part II (see also Alfven number).

$$\pi_{ma} \qquad \qquad \qquad \qquad \pi_{m\ell} = \frac{F}{i^2 \mu}$$

$$g \triangleq it \qquad f \triangleq \frac{1}{t}$$

Fluid Dynamics Laws. Inertial forces imposed by fluid motions and viscous forces resulting from internal and wall friction of the melt are governed by Newton's law of inertia and the friction law of Newtonian fluids. The convex shape of the melt's free surface suggests the influence of two more laws: gravitation and surface tension. Hence, four pi-numbers can be formulated.

Inertial force $F \triangleq ma$

$$\pi_i = \frac{F}{\rho \ell^2 v^2}$$

$$m \triangleq \rho \ell^3 \qquad a \triangleq \frac{\ell}{t^2} \qquad t \triangleq \frac{\ell}{v}$$

Viscous force

$$F \triangleq \tau \ell^2$$

$$\pi_v = \frac{F}{\rho \nu v \ell}$$

$$\tau \triangleq \rho \nu \frac{v}{\ell}$$

Gravitational
force

$$F \triangleq mg \xrightarrow{\qquad\qquad\qquad\qquad\qquad} \pi_g = \frac{F}{\rho g \ell^3}$$

$$m \triangleq \rho \ell^3$$

Surface tension
force

$$F \triangleq \sigma_s \ell \xrightarrow{\qquad\qquad\qquad\qquad} \pi_s = \frac{F}{\sigma_s \ell}$$

Note that in the formulation of π_v the use of μ for dynamic viscosity is avoided since this notation is already reserved for the magnetic permeability. Note also that σ notates electric conductivity; and σ_s, surface tension.

Gravitational forces are indicated not only by the curved shape of the surface but also by differences in density caused by differences in temperature. Consequently, a fifth pi-number expressing the influence of buoyancy must be added. The representative form of the buoyancy force derives from the law of gravitation, $F \triangleq (\rho - \rho_0) g \ell^3$, where ρ_0 is a reference density. The relation between ρ and ρ_0 can be expressed in terms of the coefficient of thermal expansion, β, as $\rho = \rho_0 (1 + \beta \theta)$. Therefore,

$$F \triangleq (\rho - \rho_0) g \ell^3 \xrightarrow{\qquad\qquad\qquad} \pi_f = \frac{F}{\rho_0 g \beta \theta \ell^3}$$

$$\rho = \rho_0 (1 + \beta \theta)$$

Heat Transfer Laws. At steady state, as much heat is generated and absorbed within the melt as is dissipated through conduction, convection, and radiation. As a result, the temperature distribution remains unaltered.

The temperature distribution at steady-state was not the object of the planned investigation, however; otherwise, four more pi-numbers derived from the laws of heat absorption, conduction, convection, and radiation would have had to be added to the large number of already derived pi-numbers. The temperature distribution was of interest only insofar as it influenced the flow pattern, and it was argued that in the temperature range of interest (about $1000^\circ C$) the flow pattern was determined primarily by the electro-magnetic field.

To prove this argument, the vertical metal velocity was measured in the de-energized furnace previously heated to 970°C. The measurements were made with a small, horizontal, pressure-sensitive plate lowered into the melt at various heights and radial locations. At no point was the metal speed larger than 1 cm/s -- an insignificant amount compared with the velocities (up to 50 cm/s) induced by the electro-magnetic field. Consequently, the influence of the heat transfer laws on the flow pattern was disregarded.

Identification of Non-Essential Laws. The low speeds of the free thermal circulation suggested negligible influence of buoyancy on the flow pattern; therefore, π_f was disregarded.

The remaining pi-numbers of π_{el}, π_{ma} or $\pi_{mb}, \pi_i , \pi_v , \pi_g$, and π_s were combined as follows.

$$\underline{R_m} = \frac{\pi_{el}}{\pi_{ma}} = \sigma \mu f \ell^2 \qquad \text{(Magnetic Reynolds No.)}$$

$$\pi_A = \frac{\pi_{mb}}{\pi_i} = \frac{\rho \ell^2 v^2}{i^2 \mu} \qquad \text{(a version of the Alfven No.)}$$

$$\pi_M = \frac{\pi_{mf}}{\pi_v} = \frac{\rho v v \ell}{i^2 \mu} \qquad \text{(a version of the Hartman No.)}$$

$$\underline{We} = \frac{\pi_s}{\pi_i} = \frac{\rho \ell v^2}{\sigma_s} \qquad \text{(Weber No.)}$$

$$\underline{Fr}^2 = \frac{\pi_g}{\pi_i} = \frac{v^2}{g \ell} \qquad \text{(Froude No.)}$$

Self-modeling eliminates the influence of the linear dimensions and of all material properties. The ensuing model rules are

$$\underline{R_m} \longrightarrow \frac{f}{f'} = 1$$

$$\pi_A \longrightarrow \frac{v}{v'} = \frac{i}{i'}$$

$$\pi_M \longrightarrow \quad \frac{v}{v'} = \left(\frac{i}{i'}\right)^2$$

$$\left.\begin{array}{l}
\underline{We} \longrightarrow \quad \dfrac{v}{v'} = 1 \\[2em]
\underline{Fr} \longrightarrow \quad \dfrac{v}{v'} = 1
\end{array}\right\} \longrightarrow \quad \dfrac{f}{f'} = 1$$

The constraints imposed on experiments by self-modeling leave little room for parameter variations. The first pi-number, R_m, representing the influence of the electric field on the flow pattern calls for an unchanged frequency. Likewise, We and Fr (representing the influences of surface tension and gravity, respectively) do not permit variation of any velocity including the rate of the time-varying field. Under these circumstances, the effects of magnetic and inertial forces (reflected in π_A), and viscous forces (reflected in π_M) cannot be investigated.

If one could prove that the effects of the electric forces, surface tension, and gravity are unimportant, so that the frequency and the velocity could be changed, then only π_A and π_M would remain prominent. Note, however, that these two pi-numbers spell out conflicting requirements. The first, repre-senting the influence of inertial and magnetic forces on the flow pattern, prescribes a linear relation between speed and electric current, $v \propto i$. The second, denoting the influence of viscous and magnetic forces, predicts a relation of $v \propto i^2$.

To study the effects of electric forces (reflected in R_m), the authors of the Sources of Information compared the melt velocities at the crucible axis at frequencies of 50 Hz and 2500 Hz. They found that the melt velocity was only twice as large at 50 Hz as it was at 2500 Hz, a frequency fifty times higher; they concluded that the influence of frequency and, hence, velocity on the circulation is weak far as R_m was concerned.

The effects of surface tension and gravity were studied by suppressing the
surface meniscus. Tests were carried out with (a) a free metal level (shaped
like a meniscus), and (b) a non-conducting ceramic cover immersed in the melt
and keeping its surface flat (Fig. 9.1). In both cases, the relation between
inductor current, i, and the velocity of the melt, v, at the crucible axis
was observed, so that the experiment served two purposes. If the current-
velocity relation were the same in both cases, then (1) gravity and surface
tension would be shown to have no influence on the circulation of the melt;
and (2) the shape of the current-velocity curve would show whether π_A or π_M
was dominating the flow pattern.

The test results, pictured in Fig. 9.2, show that (1) at low current, the
two sets of data for the free surface and the covered surface nearly coalesce,
so that the influences of gravity and surface tension can be ignored; and
(2) the relation between electric current and velocity is linear, so that π_M
and, hence, viscous influences can be disregarded. This result is not sur-
prising, for the motions of the melt were observed to be turbulent.

Thus, only inertial forces and magnetic forces turned out to be important,
and the motions of molten metal in induction furnaces can be considered to
be governed by π_A alone.

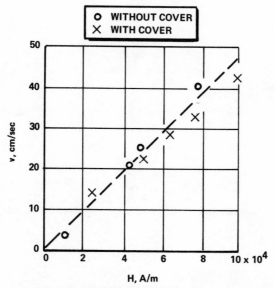

Fig. 9.2.   Relation between melt velocity, v, and magnetic field strength, H. Note that H is
proportional to the inductor current, i.

Sources of Information

L.L. Tir, "Modeling the motion of molten metal in an induction furnace," *Magnitnaya Gidrodinamika* 1, 4, 120-124 (Oct - Dec 1965).

N.K. Drakunkina and L.L. Tir, "Experimental investigation of similarity conditions in the motion of molten metal in an induction furnace," *Magnitnaya Gidrodinamika* 2, 1, 137-141 (Jan - Mar 1966).

Translation of both papers in *Magnetohydrodynamics* 1, 4, 76-78 (1965); and 2, 1, 81-84 (1966); Faraday Press, New York.

APPENDICES

APPENDICES

APPENDIX A
PROBLEMS

Problem 1: Derive the frequency scale factor of a string fastened at both ends and stretched by a force. What is the force scale factor?

Note: Disregard the effect of gravity (string weight). Observe that $\sqrt{E/\rho}$ is the speed of sound, a_s .

Perform simple model tests as pictured in Fig. A1 and verify the model rules for a length scale factor of 4. Use the same material for both model and prototype strings.

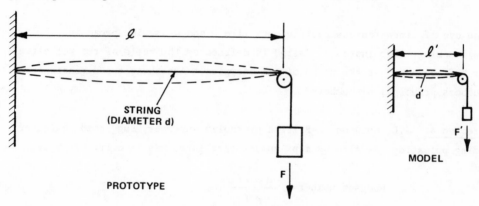

Fig. A1. String experiment.

Problem 2: Suppose the deformation of a given structure is caused by its own weight. Is it possible under these conditions to investigate the elastic response by a scale model built from the same material as that of the prototype? Neglect the influence of Poisson's ratio.

Problem 3: During the impact of two bodies, part of the initial kinetic energy is dissipated. For central impact, the loss of kinetic energy can be expressed in terms of the coefficient of restitution, e . The coefficient of restitution is defined as the ratio of two impulses:[1] e = impulse during restitution phase/impulse during deformation phase.

---

[1] I.H. Shames, *Engineering Mechanics*, Vol. II, Prentice Hall, Englewood Cliffs, N.J., 1966.

Since e is dimensionless, can it be considered a principal pi-number? Hint:
the answer depends on whether or not the relation given above can be considered a physical law applicable to a group of similar phenomena. The
coefficient of restitution depends on the initial shapes and velocities
of the bodies before impact; it also depends strongly on their materials. However, for simple shapes like spheres and for a given velocity
range, the coefficient appears to be dependent only on the materials of the
impinging bodies. Materials for which e = 0 are said to be inelastic; those
for which e = 1, perfectly elastic.

Apply similar considerations to the coefficient of friction, which is defined
as the ratio of two forces.

$$\mu = \text{tangential force/normal force.}$$

The use of dimensionless coefficients like e and μ is widespread in
engineering. For instance, "slip" is defined as the ratio of two velocities;
or "efficiency," as the ratio of two energies. Are these two dimensionless
numbers principal pi-numbers?

Problem 4:   J.L. Hodgson, a British hydraulic engineer, suggested character-
izing pulsating gas flow by a pi-number[1] that later was to carry his name:

$$\text{Hodgson number} = \frac{\Delta p \, V f}{p \, \dot{V}}$$

Figure A2 illustrates the various quantities.

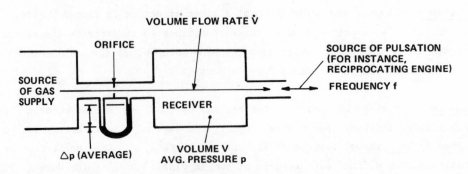

Fig. A2. Diagram of pulsating gas flow.

---

[1] J.L. Hodgson, "The laws of similarity for orifice and nozzle flows," *Trans.
ASME* 51, 22, 303-319 (Sep-Dec 1929).

Question: is the Hodgson number a principal pi-number (i.e., derived from physical laws) or is it a common pi-number?

Problem 5: Perform a simple impact model test with a pendulum as indicated in Fig. A3. Let the bob of a very long pendulum strike a heavy block resting on a smooth, level table. The bob should hit the front end of the block at the height of the center of gravity in a horizontal direction (direct, central impact). To ensure inelasticity of the impact, the front end of the block should be covered with inelastic material; for instance, lead. The momentum imparted to the block moves it a certain short distance before friction between the table and block stops the motion. Derive the model rules governing the impact and the post-impact phase, and verify the rules by a simple model experiment with two geometrically similar configurations. Observe that the collision phase is exclusively governed by inertial forces, and the post-collision phase by inertial and friction forces. The model experiments can be expanded to include oblique and excentric impacts.

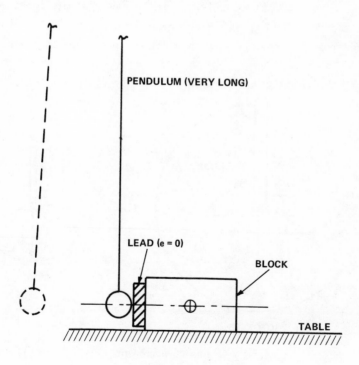

Fig. A3.   Impact pendulum experiment.

Problem 6: Figure A4 shows a simple pendulum machine used for measuring the
absorption of kinetic energy of small specimens. The pendulum strikes and
fractures the specimen midway between two supports; the energy expended is
computed from the original and the final positions of the pendulum.

The pendulum machine can be used to demonstrate the model rules governing the
collision between two objects[1] Derive the model rules for the case where one
object is rigid (pendulum); and the other one, soft or brittle (specimen).

Assume the materials of both model and prototype specimens to be the same
and neglect rate effects. Neglect also the kinetic energy of the specimen
and the stress energy of the pendulum.

What is the model rule for the pendulum mass? What is the time scale factor?
What is the scale factor of the final angle ϕ ?

Check the validity of the model rules by performing a simple model test with
two similar machines (length scale factor of the order of 4); use the same
material (for instance, wood) for both model and prototype specimens.

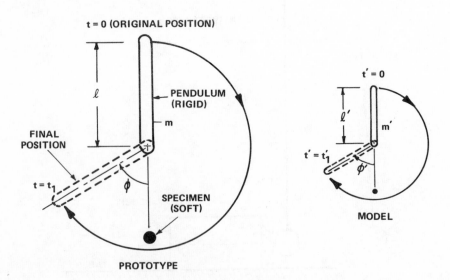

Fig. A4. Impact experiment.

---

[1]R.I. Emori, "Scale models of automobile collisions with breakaway obstacles,"
Experimental Mechanics 13, 2, 64-69 (Feb 1973).

<u>Problem 7</u>:   In a theoretical paper, Hsu[1] investigated the motions of a heavy
string hanging in a fluid and moved up and down at its support point.   The
investigation was motivated by problems arising from the usage of devices such
as drill strings in ocean engineering.   Using certain simplifications, Hsu
arrived at the following equation of motion

$$\frac{W_0}{g} \frac{\partial^2 y}{\partial t^2} + \frac{1}{2} \rho_f d_o C_d \frac{\partial y}{\partial t} \left| \frac{\partial y}{\partial t} \right| - \frac{\partial}{\partial x} \left[\left(W_1 + \frac{W_0}{g} \frac{d^2 z}{dt^2} \right) x \frac{\partial y}{\partial x} \right] = 0$$

The terms of this equation are explained in Fig. A5.

Substitute the individual terms in this equation by representative quantities
(for instance, $\partial y / \partial t$ by ℓ/t or v) and determine which physical laws were
used to derive the equation.   Which principal pi-number expresses these laws?

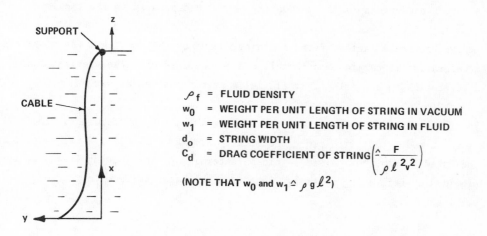

$$\rho_f = \text{FLUID DENSITY}$$
$$w_0 = \text{WEIGHT PER UNIT LENGTH OF STRING IN VACUUM}$$
$$w_1 = \text{WEIGHT PER UNIT LENGTH OF STRING IN FLUID}$$
$$d_o = \text{STRING WIDTH}$$
$$C_d = \text{DRAG COEFFICIENT OF STRING} \left(\triangleq \frac{F}{\rho \ell^2 v^2} \right)$$

(NOTE THAT w_0 and $w_1 \triangleq \rho g \ell^2$)

Fig. A5.   Hanging string with moving support.

[1] C.S. Hsu, "The response of a parametrically excited hanging string in fluid,"
J. Sound and Vibration 39, 3, 305-316 (1975).

Problem 8: The Euler number is a special form of the Newton number,
$\underline{Ne} = F/\rho \ell^2 v^2$, with the representative force F replaced by $F \triangleq \rho \ell^2$,
so that, basically, $\underline{Eu} = p/\rho v^2$. Suppose, in a gas flow problem, instead
of density ρ and velocity v, the more easily measurable quantities of
temperature, Θ , and mass flow rate, $\dot{m}$, are to be determined. Replace ρ
and v in the Euler number by Θ and $\dot{m}$ by using representative versions of the
continuity equation and the equation of state for ideal gases.

Problem 9: Fuel sloshing in missile tanks during take-off may lead to severe perturbations of the flight trajectory if one of the missile's natural
frequencies coincides with the sloshing frequency. Abramson and Ransleben[1]
performed a series of model tests to investigate the damping effects of
baffles and other suppressing devices on the sloshing frequencies and ampli-
tudes. They found sloshing to be governed by inertial and gravitational
forces; the effects of viscosity were small even with a large number of
suppression devices (such as floating cans) incorporated in the tank.

Derive the scale factor of the sloshing frequency. Assume that the missile's
acceleration is constant so that it can be combined with the gravitational
acceleration to a constant, artificial gravitational acceleration, a.

Verify the frequency scale factor (with $a = a' = g$) by performing a simple
sloshing model test with two geometrically similar bowls filled with water, as
indicated in Fig. A6. A simple aid in determining the sloshing frequency is
provided by reflecting a strong light from the water surface to the room
ceiling.

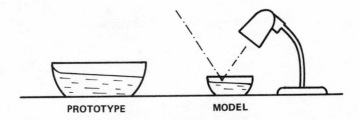

PROTOTYPE   MODEL

Fig. A6. Sloshing experiment.

[1]H.N. Abramson and G.E. Ransleben, "Simulation of fuel sloshing in missile
tanks by use of small models," *J. Am. Rocket Soc.* 30, 7, 603-612 (Jul 1969).

Problem 10: Model tests with cold air as the working fluid (instead of hot gas) are very suitable for predicting the performance characteristics of gas turbines. The flow processes are dominated by inertial and pressure forces; viscous forces are usually disregarded. Under these conditions, only the Newton number applies. Derive this number from Newton's law of inertia and express it in terms of length ℓ , rotational speed N , absolute gas temperature θ , and gas constant R .

If the scale factors of length, pressure, and temperature are given, what are the scale factors of rotational speed N (in revolutions per unit time), mass flow rate $\dot{m}$ (in mass per unit time), specific work h (in energy per unit mass), and torque?

Hint: Use the equation of state of ideal gases.

Problem 11: When an aircraft approaches the speed of sound, randomly fluctuating airloads may occur, causing the wings to vibrate through a wide range of frequencies. This phenomenon is called transonic buffeting.

Hollingsworth and Cohen[1] compared the onsets of buffeting of a McDonnell Douglas F-4 airplane and its 1/20 scale model (tested in various wind tunnels). The onset of buffeting was indicated by a rather sharp increase in the rms value of the wing-tip accelerations.

It was argued that the onset of buffeting is governed by the oscillatory characteristics of the wings (particularly by their first bending frequency), and by the inertia and elasticity (compressibility) of the air flow. Hence, four physical laws were applied representing the

$$\text{elastic wing forces} \qquad F \triangleq k \ell ,$$

$$\text{inertial wing forces} \qquad F \triangleq m v^2 / \ell ,$$

$$\text{inertial forces of air} \qquad F \triangleq \rho \ell^2 v^2 ,$$

$$\text{adiabatic compression of air} \qquad F \triangleq \rho a_s^2 \ell^2 / \gamma ,$$

---

[1] E.G. Hollingsworth and M. Cohen, "Comparison of wind tunnel and flight test techniques for determining transonic buffet characteristics on the McDonnell Douglas F-4 airplane," AIAA Paper 70-584 presented at AIAA 5th Aerodynamic Testing Conf., May 1970.

where k is the wing's spring rate (first symmetrical bending mode); m is
the characteristic wing mass (also representing the mass distribution of the
wing); and ρ is the air density. Derive the scale factors for speed, frequency, and mass. The same air was used for model and prototype. Remember
that the bending frequency can be represented by $\sqrt{k/m}$. If the onset of
prototype buffeting is observed at a given angle of attack and a given speed,
determine the corresponding angle and speed of the model.

Problem 12: Any direct measurements of the elastic properties of red blood
cells are frustrated by their minute size (a few microns). Therefore,
indirect methods were tried, but their results were doubtful because they
involved serious cell distortions to the point of cell damage. Hoeber and
Hochmuth[1] developed a new indirect method in which the relaxation time of
red blood cells was compared with that of much larger artificial cells of
known elastic properties. In both cases, strain and stress were kept at low,
natural levels. By using proper scaling methods, they could estimate the
elasticity of the original cells.

The Hoeber-Hochmuth experiment exemplifies the case where the model is larger
than the prototype. The authors designed a scale model 5000 times larger
than the prototype. Since they considered a red blood cell a thin-walled,
liquid filled biconcave shell, the model was manufactured from large rubber
cells filled with silicone oil. Both blood cells and rubber cells, when
suspended in fluid and deformed slightly, were assumed to exhibit linear
elasticity. Since, after deformation, restoration to the original form
takes place slowly, inertial forces were excluded so that only elastic and
viscous forces needed be modeled. Under these circumstances, what are the
two principal pi-numbers governing the relaxation after deformation?

In the experiment, small deformations were imposed upon a cell by sucking it
into a pipette whose diameter was slightly smaller than that of the cell.
The cell was then expelled from the pipette into a fluid of known viscosity,
and the time required by the cell to return to its relaxed form was measured.

---

[1] T.W. Hoeber and R.M. Hochmuth, "Measurement of red cell modulus of elasticity
by in-vitro and model cell experiments," *Trans. ASME, J. Basic Eng.* 92, D3
604-609 (Sep 1970).

If the average relaxation time of the (large) model is 4.0 sec, and of the blood cells, 0.24 sec, what is Young's modulus of elasticity of the red blood cell? Assume Young's modulus of the model to be $E' = 9.2 \times 10^5$ N/m^2. Assume also that the viscosity of the prototype fluid was 1.0 cp and that of the model, 230 cp.

If the ratio of the viscosities of the intracellular and the extracellular fluids of the blood is 4.1, what is the desired corresponding ratio of the model?

Problem 13: The Horton number, designated by Strahler[1] to honor Robert E. Horton, an American hydraulic engineer (1875-1945), is claimed to indicate the erosion intensity of land slopes by water runoff. The number derives from an empirical relation among the soil mass removed by runoff water, m_s; the eroding force per unit surface, t_e ; the area exposed to erosion, A ; and the time, t .

$$m_s = k_e A t_e t$$

The erosion proportionality factor, k_e , is a property of the land surface: for silt or sand, k_e is high; for rock or heavily vegetated soil, it is low. The factor k_e is combined with the so-called runoff intensity, Q , to form the Horton number.

$$\underline{Ho} = Q k_e$$

The runoff intensity is defined as the volume rate of surface water flow per unit area of cross section; in other words, Q is the average speed of the runoff water, v . Derive the Horton number from Horton's empirical relation for m_s and from Newton's law of inertia. Hint: t_e can be expressed representatively as F/A , and Newton's law as $F \triangleq m v/t$.

---

[1]A.N. Strahler, "Dimensional analysis applied to fluvially eroded landforms," *Bull. Geological Soc. Am.* 69, 279-300 (Mar 1968).

Problem 14: In Chapter 3, a seepage model test is described. Perform this test as shown in Fig. 24 by using sand of different grain sizes, with the depth of the sand layer made proportional to the grain size. Since the flow is essentially one-dimensional, the same tube can be used for all experiments. Measure the flow velocity by timing the water level and plot the results in form of (ℓv) versus (z / v^2) on log-log paper. The test results should fall on a straight line with a slope of 45 degrees.

Problem 15: The gross dynamic response of a ship to waves is governed principally by the inertial and gravitational forces of both ship and water. Hence, Newton's laws of inertia and gravitation apply. Since the ship can be treated as a rigid mass, the two laws may be expressed separately for ship and water. In representative terms,

for the ship

inertial force $\qquad F \triangleq m v^2 / \ell$

gravitational force $\qquad F \triangleq mg$

for the water

inertial force $\qquad F \triangleq \rho \ell^2 v^2$

gravitational force $\qquad F \triangleq \rho g \ell^3$

Derive the four principal pi-numbers that evolve from these four representative expressions. Combine them into three new pi-numbers by eliminating the representative force, and show that two of the new numbers are identical with the Froude number.

The Froude number does not contain the density, ρ. Does it follow that the density scale factor can be selected arbitrarily?

Moreau[1] describes model and full scale tests performed on a Belgian cargo liner. During full scale sea tests, the heave and pitch motions of the ship and the power spectra of the waves were recorded. The model, made of Fiberglas with a length scale of 1:100, was towed in a basin with water as test fluid. With $\rho^* = 1$ and $\ell^* = 100$, what are the scale factors of the ship mass and of the radius of gyration around the center of gravity?

---

[1]MIT, Dept. Naval Arch. and Marine Eng., *Ship Model Tests in Irregular Seas and Correlation with Full-Scale Trials*, by W.J. Moreau, Rep. 65-12, Cambridge, Mass, Dec 1965.

The model waves were scaled from the sea wave spectrum measured in sea tests. Each sinusoidal component of the sea spectrum was scaled and generated separately, and all components were superimposed randomly. What is the wave frequency scale factor?

Problem 16: The resistance of soft soil to vertical deformation by a wheel or track can be described by an empirical pressure-sinkage relationship suggested by Bekker[1]

$$p = k_c + k_\phi z^n$$

where p is the vertical mean pressure across the penetration gear surface, z is the vertical penetration depth, and k_c, k_ϕ, and n are soil constants. For sand, k_c is relatively small and can be neglected. What is the vehicle weight scale factor if the same sand is used for model and prototype? Assume a length scale factor of 4 and an exponent n of 1.1.

Problem 17: In Case Study 5, model rules were developed for an off-road vehicle moving over sand. Suppose the Marsh Buggy and its model are moving over clay soil. What is the weight scale factor?

At equal slip values of model and prototype, what is the expected scale factor of drawbar pull?

If model and prototype are moving under their own power, what is the expected scale factor of power (at the wheels) at equal slip values?

Performance data of model and prototype should not be compared at unequal slip values. Why not?

Problem 18: Figure 1 shows the scale model of a lunar rover.[2] The length scale factor is 6. Derive this scale factor by using the following assumptions:

---

[1] M.G. Bekker, *Off-The-Road-Locomotion*, University of Michigan Press, Ann Arbor, Mi, 1960.

[2] See also: D. Schuring, "Scale model testing of land vehicles in a simulated low gravity field," *SAE Trans.* 75, 699-705 (1967).

- lunar soil is cohesionless,

- the rover moves at low speed so that inertial forces of the soil
  need not be considered (inertial forces of the rover must be
  considered, however),

- the rover's wire mesh wheels and suspension springs are elastic
  and have negligible internal friction,

- the reaction force of the rover's shock absorbers is linearly
  dependent on speed; i.e., $F \triangleq k v$ (k = damping rate),

- the unsprung mass of the rover is rigid,

- both model and prototype are made from the same material so that
  $E = E'$ and $\rho = \rho'$.

What are the scale factors of speed, force, power, and damping rate?

Problem 19: During the slow descent of a space vehicle landing on a planetary
surface, rocket blast can severely erode the soil beneath the craft and hence
degrade its landing stability. To study the cratering process of a lunar
landing, Land and Schott[1] performed model experiments in a large vacuum
cylinder. The soil was assumed to be cohesionless; it was therefore represented by small glass beads. The rocket blast was generated by an overhead
nozzle with cold helium as exhaust gas. The nozzle was first lowered and then
held at terminal height to simulate hovering. The soil erosion process was
recorded continuously by an X-ray apparatus.

Find the scale factors for length, speed, and thrust. Remember that
g_{earth} = 6 g_{moon}. Use the three pi-numbers π_i , π_g , and π_μ derived in
Case Study 5 for modeling cohesionless soil. Assume that the transfer of
thermal energy has negligible influence on erosion so that besides the soil
forces only the kinetic and internal gas energies must be modeled. Note that
the kinetic gas energy is already represented by the soil pi-number for inertia
effects, and that the internal gas energy can be expressed by the first law
of thermodynamics for an ideal gas.

---

[1]N.S. Land and H.F. Schott, "Scaled lunar module jet erosion experiments,"
NASA TN-D-5051, Washington, D.C., Apr 1969.

The temperature of the prototype gas at the nozzle exit was known to be 672 K. The temperature of the model gas was 85 K, and the ratio of c_v / c_v' was 0.41. The densities of the two gases were approximately equal, and so were (presumably) the densities of the model and the planetary soil. The coefficients of internal friction of both soils were also assumed to be equal.

Problem 20: To understand the basic mechanism of soil tilling, Krause[1] moved geometrically similar, narrow blades at constant speed horizontally under the surface of dry, loose sand and measured the resistance, F_H, as indicated in Fig. A7. He found that for the same Froude numbers, i.e., for gh / v^2 = const, the dimensionless resistances, $F_H \, g^2 / \rho \, v^6$, would be the same for all geometrically similar tools run at the same dimensionless depth, t/h.

Derive this result from the three pi-numbers developed for sand in Case Study 5. (Assume that the same sand is used for all geometrically similar tools, so that $\mu = \mu'$.) How did Krause arrive at the rather complex expression for the dimensionless resistance? Could he have used a simpler dimensionless expression instead, such as $F_H / \rho g \ell^3$? Why are force comparisons for similar tool configurations restricted to equal dimensionless depths?

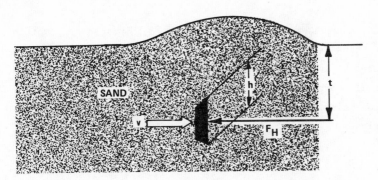

Fig. A7. Tillage tool moving from left to right.

Problem 21: In Case Study 7.1 "Pulsating fires and pulsating stars," a simple model experiment is described with ethanol fuel burned in shallow metal pans. Perform a similar experiment.

---

[1] R. Krause, "The most important phenomena subsoiling in dry sand," *J. Terramechanics* 12, 3/4, 119-130 (Dec 1975).

Problem 22: Heat transfer between a fluid at one temperature and a solid
wall at a different temperature is usually described by a formula suggested
by Newton,

$$\dot{Q} = hA \, \Delta\theta ,$$

where $\dot{Q}$ is the rate of the heat transferred, A is the area of the wall,
and $\Delta\theta$ is the temperature difference between the wall and the main body of
the fluid. The heat transfer coefficient, h , is not a material constant;
it depends on many factors such as the geometry of the wall, the physical
properties of the fluid, fluid velocity, and temperature. Hence, Newton's
equation is a definition (of h) rather than a physical law.

Suppose a model test must be designed involving heat transfer between a
gas and a wall. Since the functional relations between h and temperature,
speed, wall geometry, and fluid properties are not known, it is mandatory
that the model's and the prototype's coefficients of heat transfer be the
same. This can be achieved by using the same fluid for model and prototype
and by maintaining (besides geometric similarity) the same corresponding
temperatures and fluid speeds. Under these rather stringent circumstances,
what is the scale factor of energy in terms of the length scale factor?

Problem 23: Suppose a closed, well insulated, nonexpanding container of
arbitrary shape is filled partially with a liquid of given pressure and
temperature. A known amount of heat is then supplied to the liquid, raising
its internal energy and, with it, the temperature and the pressure. The
process governing the temperature and pressure rise can be described representatively by

$$E \triangleq c_v \rho \, \ell^3 \theta$$

and

$$E \triangleq \lambda \rho \, \ell^3$$

where λ is the specific heat of vaporization. Since the material constants
in these relations are functions of temperature, it is mandatory to employ
a temperature scale factor of unity and to use the same fluid for model and
prototype. Under these circumstances, what is the pressure scale factor?
How does one enforce a temperature scale factor of unity?

Problem 24: During braking, the kinetic energy of a decelerated vehicle is converted into heat energy in the brakes. The heat energy is conducted from the surface of the brake lining into the brake, where part of it is stored and part of it convected and radiated. However, since stopping an automobile usually takes only a few seconds, not much heat is dissipated during the stopping period. For modeling the temperature rise in a brake during braking, then, the energies of interest are the kinetic energy of the vehicle, the heat energy stored in the brake, and the heat energy conducted away from the lining surface into the brake. Hence, three laws apply: Newton's law of inertia (expressed in form of kinetic energy), the law for heat capacity, and the law for heat conduction (Fourier). In representative terms,

Kinetic energy	$E \triangleq \rho \ell^5 / t^2$
Stored heat energy	$E \triangleq c \rho \ell^3 \Delta \theta$
Conducted heat energy	$E \triangleq k \ell \Delta \theta t$

Bäse[1] performed model tests with geometrically similar brakes made from the same materials. Show that under these circumstances, $\Delta\theta^* \ell^{*2} = 1$ and $t^* / \ell^{*2} = 1$.

In the test setup, the brake was attached to a flywheel, and both brake and flywheel were brought (by a power source) to an initial energy level, E_o, at an initial speed, n_o. What are the scale factors of E_o and n_o in terms of the length scale factor?

When the flywheel reached the desired speed and energy levels, it was disconnected from the power source and then braked at constant torque (to simulate constant vehicle deceleration) until the flywheel came to a complete stop. What is the torque scale factor in terms of the length scale factor? What are the expected scale factors of peak temperature and stopping time?

---

[1] W. Bäse, "Modellaehnlichkeitsversuche an Trommel- und Scheibenbremsen," (Model tests on drum and disk brakes), *Deutsche Kraftfahrtforschung und Strassenverkehrstechnik* 198 (1968), VDI Verlag, Duesseldorf.

Problem 25:   In Case Study 1, modeling of explosive forming was shown to be
governed by two principal pi-numbers (derived from Newton's law of inertia
and the stress-strain relations of the materials involved) which, with the
same materials used in both model and prototype, resulted in the two model
rules for explosive energy and speed, $E^* = \ell^{*3}$ and $v^* = 1$.

Suppose that instead of a chemical explosive, a thin wire connected to a
large capacitor is used, as indicated in Fig. A 8. The electrical energy
stored in the capacitor would, when discharged, "explode" the wire and
generate a shock wave like a chemical explosive. Martin[1] used geometrically
similar devices of this kind to study the effects of different loadings in
small-scale explosion-forming experiments. The same materials for the
exploding wires (copper), the diaphragms to be deformed (1100-0 aluminum),
and the fluids (water), were used. The scale factor for the capacitance
was $c^* = \ell^{*2}$, and for the voltage, $u^* = \sqrt{\ell^*}$. Derive these two relations
from the two model rules stated above and from the representative relations
for electric energy, $E \triangleq uit$, and capacitance, $C \triangleq \dfrac{it}{u}$.   Also use Ohm's law
$(i \triangleq u/R)$ and the fact that the electric conductivities of both model and
prototype wires are the same.

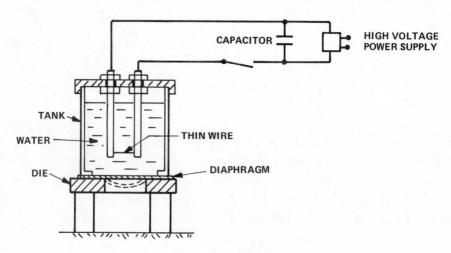

Fig. A8.   Test fixture with electrically charged thin wire as "explosive."

---

[1]C.W. Martin, "Similitude in exploding wires," *Experimental Mechanics* 15, 2
49-54 (Feb 1975).

Problem 26: The concepts of physical similarity have been extended by
Rashevsky[1] and others[2] to include the structural makeup of organisms. Based
on the observation that all mammals are approximately similar in form and of
the same overall mass density, it is claimed that the surface area, A, and
mass, m, must assume a representative relation of the kind $A \triangleq m^n$. What is
the exponent n? If, likewise, the lungs of all mammals are geometrically
similar and of the same mass density, what is the representative relation
between lung volume, V_L, and mass, m?

The metabolic energy rate of a mammal is proportional to the rate of oxygen
supply, which in turn is proportional to the product of $f_L V_L$, where f_L is
the frequency of respiration. Now, it is well known that small mammals
breathe faster than larger ones. Rubens suggested that the metabolic rate
is proportional to an animal's surface area; i.e., $f_L V_L$ prop A. With this
"law" and the representative relations derived earlier, what is the expected
relation between the frequency of respiration, f_L, and the mammal's mass, m?
Check the relation by plotting the following data:

animal	f_L $\min^{-1}$	m kg
rat	115	0.5
rabbit	60	3
cat	30	5
dog	25	25
sheep	15	40
bear	11	300
horse	10	600

Problem 27: The direction, time history, and amplitude of early sound
reflections are important attributes of the acoustical quality of a music
hall. If delayed by more than 70 ms and if sufficiently loud, reflections are
generally heard as echos -- a very undesirable effect.

[1] N. Rashevsky, *Mathematical Biophysics*, Part IV (The organism as a whole, and
the organic world as a whole), University of Chicago Press, Ill, 1948,
pp. 569-615.

[2] W.R. Stahl, "The analysis of biological similarity," *Advances in Biological
and Medical Physics*, 9, 355-464 (1963).

The echo sensitivity of an auditorium is usually measured by releasing a spark of given energy on the stage and recording its echo at various locations in the hall. Suppose echo measurements are to be made for the model of a concert hall. Determine the scale factor for spark energy. Pertinent pi-numbers are given in Case Study 8. What is the sound pressure scale factor?

APPENDIX B

CATALOG OF PRINCIPAL PI-NUMBERS

Principal pi-numbers are based on not more than two dozen laws. Nevertheless, more than one hundred principal pi-numbers have evolved over the past century or so. The reason for this abundance is of course the almost infinite number of possible combinations of laws. Some principal pi-numbers, such as the Newton and the Nusselt numbers, contain but one physical law. Others, such as the Froude and the Reynolds numbers, are constituted from two laws or, like the Grashof number, from three. The Prandtl number is one of the few assembled from four laws.[1]

To inventory all principal pi-numbers that have ever been suggested would be cumbersome indeed.[2] Some, mentioned a few times in the literature, have not been generally accepted; others, occurring in specialized fields, are used only infrequently. Still others have been dropped from use or are applied only in certain countries. Several principal pi-numbers are known by more than one name; a few share their names with others.

We have tried to establish a midcourse by presenting principal pi-numbers frequently used in recent literature plus a few others less frequently used but with an interesting structure. We will present them together with their constituting laws and other important information in the form of a table, starting with the Alfven number and ending with the Z-number. The tabular form, chosen for brevity, calls for a few explanations.

---

[1] Two of the pi-numbers presented in the following catalog, the Damköhler group and the Thoma number, are based on no law at all but on definitions. But even though they are not principal pi-numbers, we list them because they are quoted frequently in the literature.

[2] An excellent and perhaps nearly complete collection of more than 180 named pi-numbers has been compiled by Norman S. Land in *A Compilation of Non-Dimensional Numbers*, NASA SP-274, Natl. Aeronautics and Space Administration, Washington, D.C., 1972.

The table is arranged in eight columns. The first four list those data
commonly found in tables of this kind -- name and short notation of the
pi-numbers, dimensionless product of constituting quantities, and major
fields of application. The short notations used here are not generally
accepted; workers in the field of scale modeling have never agreed on a
standard nomenclature. Some authors, objecting to the usage of two letters
such as Re for Reynolds number, suggest that they be enclosed in parentheses,
or written in a bold or italic typeface. Others promote the introduction of
the letter N for all dimensionless numbers, with the pi-number indicated in a
subscript, such as N_{Re} for Reynolds number. We have adopted the more commonly used italic-type two-letter notation, such as Re for Reynolds number or
Ma for Mach number.

In columns (5) through (7) we have attempted to indicate the fundamental
relation between principal pi-number and physical laws. Column (5) lists the
constituting physical laws and the major equations and definitions incorporated
into the pi-number. In column (6) the laws, equations, and definitions listed
in column (5) are converted into principal pi-numbers. In most cases, the
reader should find it easy to derive column (6) from column (5). The few
cases involving more than simple transactions are explained in Notes.

Column (7) specifies the basic definitions of the named pi-numbers in terms
of the constituting pi-numbers given in column (6). By substituting the constitutive pi-numbers, column (6), into the basic definitions given in column
(7), one arrives at the dimensionless products presented in column (3).

Finally, column (8) offers remarks, references, and some biographical data
about the scientists or engineers whose names were given to the pi-numbers.

Most named pi-numbers use their constituting laws to the fullest extent
possible. For instance, the Cauchy, Froude, and Newton numbers realize
Newton's law of inertia in all its manifestations -- as inertial, centrifugal,
and Coriolis forces (in any direction) as well as kinetic energy, change of
momentum, or moment of momentum. A few pi-numbers, however, make only
restricted use of their constituting laws. The Ekman number, for instance,
utilizes only the Coriolis version of Newton's law; in the Cauchy number,
the effects of Poisson's ratio -- a major parameter in Hooke's law -- are

The derivation of principal pi-numbers from physical laws as indicated in columns (5) through (7) is a step that, we believe, will promote the correct interpretation and application of pi-numbers; each principal pi-number impresses upon the model the very same laws from which it is derived. This interpretation prevents the employment of other, more mechanical, methods which may lead to unnecessarily complex combinations of many laws into one pi-number, as evidenced for example by the M-number or the McAdams group. (The M-number encompasses four laws, the McAdams group, five.) We contend that principal pi-numbers derived from more than two laws are rarely useful unless their combination results in a simple product of nothing but physical constants, such as the Prandtl number and the Schmidt number. Both numbers, although composed of, respectively, four and three laws, can themselves be considered physical constants, to be kept equal for both model and prototype. But regardless of these occasional advantages, one must always keep in mind that a principal pi-number derived from more than two laws does not eliminate the need for additional principal pi-numbers: n governing laws require n independent principal pi-numbers regardless of whether they are assembled from one, two, or more laws.

It is only with insight into the very nature of principal pi-numbers that a catalog of dimensionless numbers can be used with advantage.

(1) Name	(2) Symbol	(3) Product Derived from (6) and (7)	(4) Fields of Application	(5) Constituting Laws, Equations, and Definitions
Alfven	$\underline{AL}$	$\dfrac{v}{H}\sqrt{\dfrac{\rho}{\mu}}$	Magnetohydro- dynamics	• Lorentz's equation for magnetic force $\vec{F} = q\,\vec{v} \times \vec{B}$ • Maxwell's equation of magnetic induction $\vec{\nabla} \times \vec{H} = \vec{j}\quad (\vec{D} \approx 0)$ • Relation between $\vec{B}$ and $\vec{H}$ $\vec{B} = \mu\,\vec{H}$ • Definitions $\vec{j} = \rho_V\,\vec{v}\,;\quad i = \int_A \vec{j}\,d\vec{A}$ • Newton's law of inertia $\vec{F} = m\,\vec{a}$
Bingham	$\underline{Bm}$	$\dfrac{\tau_0\,\ell}{\mu_B\,v}$	Slow flow of rigid- viscoplastic material	• Constitutive equation of Bingham material $\tau_{yx} = \tau_0 + \mu_B\,\dfrac{\partial v_x}{\partial y}\;$; etc. τ_0 = yield stress μ_B = Bingham viscosity
Biot	$\underline{Bi}$	In form identical with Nusselt number (Note 2)		
Bond	$\underline{Bo}$ $= \dfrac{We}{(\underline{Fr})^2}$	$\dfrac{\rho g\,\ell^2}{\sigma}$	Sloshing under low gravity; liquid atomization	• Potential energy (law of gravitation) $E = mgh$ • Surface energy of liquids $E = \sigma A$

Named Pi-Numbers, their Definitions, Fields of Application, and Constituting Laws

(6) Constituting Pi-Numbers Derived from (5)	(7) Basic Definition of Named Pi-Number	(8) Biographical Data, References, and Remarks
$\pi_{m1} = \dfrac{F}{H^2 \ell^2 \mu}$ (Note 1) $\pi_i = \dfrac{F}{\rho \ell^2 v^2}$	$\underline{AL} = \sqrt{\dfrac{\pi_{m1}}{\pi_i}}$	Hannes Olof Alfvén; Swedish physicist; born 1908
$\pi_o = \dfrac{\tau}{\tau_o}$ $\pi_{VB} = \dfrac{\tau \ell}{\mu_B v}$	$\underline{Bm} = \dfrac{\pi_{VB}}{\pi_o}$	Eugene Cook Bingham; American chemist; 1878-1945
In form identical with Nusselt number (Note 2)		Jean-Baptist Biot; French mathematician, astronomer, and physicist; 1774-1862
$\pi_g = \dfrac{E}{\rho g \ell^4}$ $\pi_s = \dfrac{E}{\sigma \ell^2}$	$\underline{Bo} = \dfrac{\pi_s}{\pi_g}$	N.W. Bond, *Phil. Mag.* 5, 7th Series, 30 (Mar 1928), 794-800

(1) Name	(2) Symbol	(3) Product Derived from (6) and (7)	(4) Fields of Application	(5) Constituting Laws, Equations, and Definitions
Cauchy	Ca	$\dfrac{\rho v^2}{E}$	Vibration of elastic systems	• Hooke's law of isotropic solids $\epsilon_x = \dfrac{1}{E}\left[\sigma_x - \nu(\sigma_y + \sigma_z)\right]$, etc. $\gamma_{xy} = \dfrac{2}{E}(1+\nu)\,t_{xy}$; etc. • Newton's law of inertia $\vec{F} = m\vec{a}$
Cavitation[1]	σ	$2\,\dfrac{\Delta p}{\rho v^2}$ Δp = Difference between static pressure and vapor pressure	Cavitation	Same as Euler number
Condensation	Cv	$\dfrac{\ell^3 \rho^2 g\, \lambda}{\mu\, k\, \Delta\theta}$ (1)	Heat transfer from condensating vapor to cold surface	• Newton's law of gravitation $\vec{F} = m\vec{g}$ • Shear stress of Newtonian fluid $t_{xy} = \mu\left(\dfrac{\partial v_x}{\partial y} + \dfrac{\partial v_y}{\partial x}\right)$, etc.
	Co $= Nu\left(\dfrac{Fr}{Re}\right)^{2/3}$	$\dfrac{h}{k}\sqrt[3]{\dfrac{v^2}{g}}$ (2)		• Fourier's law of heat conduction $g = -k$ grad θ • Definition of latent heat $Q = \lambda m$ • Newton's law of inertia $\vec{F} = m\vec{a}$ • Definition of heat flux between fluid and surface $g = h(\theta_{surf} - \theta_{fluid})$
Damköhler	D_I	$\dfrac{U\ell}{\rho v}$ (1)	Chemical reactions in flowing gaseous mixture; combustion	• Definition of mass released through chemical reaction $m_U = UVt$

Named Pi-Numbers, their Definitions, Fields of Application, and Constituting Laws (Cont.)

(6) Constituting Pi-Numbers Derived from (5)	(7) Basic Definition of Named Pi-Number	(8) Biographical Data, References, and Remarks
$\pi_e = \dfrac{F}{E \ell^2}$ $\pi_i = \dfrac{F}{\rho \ell^2 v^2}$	$\underline{Ca} = \dfrac{\pi_e}{\pi_i}$ (Influence of v neglected)	Baron Augustin Louis Cauchy; French mathematician; 1789-1857
Same as Euler number		[1]Special form of Euler number
$\pi_g = \dfrac{F}{\rho g \ell^3}$ $\pi_v = \dfrac{F}{\mu \ell v}$ $\pi_k = \dfrac{Q}{k \ell \Delta \theta t}$ $\pi_\lambda = \dfrac{Q}{\lambda \rho \ell^3}$ $\pi_i = \dfrac{F}{\rho \ell^2 v^2}$ $\pi_h = \dfrac{Q}{h \ell^2 \Delta \theta t}$	$\underline{Cv} = \dfrac{\pi_v\, \pi_k}{\pi_g\, \pi_\lambda}$ $\underline{Co} = \dfrac{\pi_k}{\pi_h} \sqrt[3]{\dfrac{\pi_g\, \pi_i}{\pi_v^2}}$	[1]R.E. Johnstone and M.W. Thring, *Pilot Plants, Models, and Scale-Up* *Methods in Chemical Engineering,* McGraw-Hill, New York, 1957 [2]W.H. Adams, *Heat Transmission,* McGraw-Hill, New York, 1954 $\underline{Co}$ can be considered a dimensionless surface coefficient of heat transfer.
$\pi_u = \dfrac{m}{u \ell^3 t} = \dfrac{\rho}{ut} = \dfrac{\rho v}{u \ell}$	$\underline{D}_I = \dfrac{1}{\pi_u}$ [1]	Gerhard Damköhler; German chemist; 1908-1944 [1] $\underline{D}_I$ is not a principal pi-number, it is based on definitions rather than on laws.

(1) Name	(2) Symbol	(3) Product Derived from (6) and (7)	(4) Fields of Application	(5) Constituting Laws, Equations, and Definitions
Damköhler (Cont'd)	$\underline{D}_{II}$ $= \underline{D}_I\,\underline{Pe}^*$	$\dfrac{U\ell^2}{D\rho}$		• Fick's law of diffusion $\quad \dot{j} = -\,D\;\text{grad}\;\rho$ • Law of heat capacity $\quad Q = c_p\,m\,\Delta\theta$ • Definition of heat released through chemical reaction $\quad Q = g_e\,m$ • Fourier's law of heat conduction $\quad g = -k\;\text{grad}\;\theta$
	$\underline{D}_{III}$	$\dfrac{g_e\,U\ell}{c_p\,v\rho\,\Delta\theta}$		
	$\underline{D}_{IV}$ $= \dfrac{\underline{D}_{III}}{\underline{Fo}}$	$\dfrac{g_e\,U\ell^2}{k\,\Delta\theta}$		
Darcy (1) friction factor	$\underline{f}$ Darcy	$2\,\dfrac{\Delta p}{\rho v^2}\,\dfrac{d}{\ell}$ d = pipe diam. ℓ = pipe length	Pressure loss in pipe flow	Same as Euler number
Dean[1]	$\underline{De}$	$\dfrac{rv}{v}\sqrt{\dfrac{r}{R}}$	Flow in curved pipes or channels	Same as Reynolds number. Inertial forces are separated into centrifugal and "axial flow" components. Hence, two representative lengths: R = radius of pipe or channel curvature r = length other than R (for instance, pipe radius) (See section "Segmented modeling," Part I)
Drag[1] coefficient	$\underline{C}_D$	$\dfrac{2\,F_D}{\rho v^2 A}$ (2)	Resistance of bodies moving through fluid	See Newton number
Eckert	$\underline{Ec}$	$\dfrac{v^2}{c_p\,\Delta\theta}$	Forced heat convection	• Law of heat capacity $\quad Q = c_p\,m\,\Delta\theta$

Named Pi-Numbers, their Definitions, Fields of Application, and Constituting Laws (Cont.)

(6) Constituting Pi-Numbers Derived from (5)	(7) Basic Definition of Named Pi-Number	(8) Biographical Data, References, and Remarks
$\pi_D = \dfrac{m}{D\rho\ell t} = \dfrac{\ell^2}{Dt}$ $\pi_c = \dfrac{Q}{c_p\rho\ell^3\Delta\theta}$ $\pi_E = \dfrac{Q}{g_e\rho\ell^3}$ $\pi_k = \dfrac{Q}{k\ell\,\Delta\theta t}$	$\underline{D}_{\mathrm{I\!I}} = \dfrac{\pi_D}{\pi_U}$ $\underline{D}_{\mathrm{I\!I\!I}} = \dfrac{\pi_c}{\pi_E\,\pi_U}$ $\underline{D}_{\mathrm{I\!V}} = \dfrac{\pi_k}{\pi_E\,\pi_U}$	
Same as Euler number		Henry Darcy; French hydraulic engineer; 1803-1858 [1]Special form of Euler number. The factors "2" and "α/ℓ" have no physical significance, only historical reasons.
Same as Reynolds number. Inertial forces are separated into centrifugal and "axial flow" components. Hence, two representative lengths: R = radius of pipe or channel curvature r = length other than R (for instance, pipe radius) (See Section "Segmented Modeling," Part I)		William Reginald Dean; British mathematician and physicist; 1896-1973 [1]Special form of Reynolds number
$\pi_i = \dfrac{F}{\rho A v^2}$ A = cross-sectional area	$c_{\underline{D}} = 2\,\pi_i$ [2] with $F = F_D$ (drag)	[1]Special form of Newton number [2]The factor "2" has no physical significance
$\pi_c = \dfrac{Q}{c_p\rho\ell^3\Delta\theta}$	$E_{\underline{c}} = \dfrac{\pi_c}{\pi_i}$ with $Q \triangleq E$	Ernst Rudolf Georg Eckert; German/US thermodynamicist; born 1904

(1) Name	(2) Symbol	(3) Product Derived from (6) and (7)	(4) Fields of Application	(5) Constituting Laws, Equations, and Definitions
Eckert (Cont'd)				• Kinetic energy (Newton's law of inertia) $E = \dfrac{m}{2} v^2$
Ekman[1]	Ek	$\dfrac{\nu}{\ell^2 \omega}$	Meteorology	Same as Reynolds number. However, only the Coriolis component of the inertial force is considered. • Inertial force = Coriolis force $\vec{F} = 2m\,\vec{\omega} \times \vec{v}$
Euler[1]	Eu	$2\,\dfrac{p}{\rho v^2}$ (1)	Fluid friction in conduits	• Newton's law of inertia $\vec{F} = m\vec{a}$
Fanning[1] friction factor	f_{fanning} $= f_{\text{darcy}}\,\dfrac{r_h}{d}$	$2\,\dfrac{\Delta p}{\rho v^2}\,\dfrac{r_h}{\ell}$ r_h = hydraulic radius[2]	Pressure loss in pipe flow	Same as Darcy friction factor
Fourier	Fo	$\dfrac{\alpha}{\ell v}$	Heat conduction	• Fourier's law of heat conduction $g = -k\ \text{grad}\ \theta$ • Law of heat capacity $Q = c_p\, m\, \Delta\theta$
Froude	Fr	$\dfrac{v}{\sqrt{g\ell}}$	Motions of accelerated masses in gravitational field; surface waves	• Newton's law of inertia $\vec{F} = m\vec{a}$ • Newton's law of gravitation $\vec{F} = m\vec{g}$

Named Pi-Numbers, their Definitions, Fields of Application, and Constituting Laws (Cont.)

(6) Constituting Pi-Numbers Derived from (5)	(7) Basic Definition of Named Pi-Number	(8) Biographical Data, References, and Remarks
$\pi_i = \dfrac{E}{\rho l^3 v^2}$		
$\pi_\omega = \dfrac{F}{\rho l^3 \omega v}$ $\pi_V = \dfrac{F}{\mu l v}$; (see $\underline{Re}$)	$\underline{Ek} = \dfrac{\pi_\omega}{\pi_V}$	Vagn Walfrid Ekman; Swedish oceanographer and hydrodynamicist; 1874-1954 [1]Special form of Reynolds number
$\pi_i = \dfrac{F}{\rho l^2 v^2}$	$\underline{Eu} = 2\pi_i$ [1] with $F \triangleq \rho l^2$	Leonard Euler, Swiss mathematician and physicist; 1707-1783 [1]Special form of Newton number. The factor "2" has no physical significance (the term $\rho v^2/2 \equiv p_d$ is known as dynamic pressure)
Same as Darcy friction factor		John Thomas Fanning; American civil engineer; 1837-1911. [1]Special form of Euler number [2]See section "Segmented Modeling"
$\pi_k = \dfrac{Q}{k l \Delta\theta t}$ $\pi_c = \dfrac{Q}{c_p \rho l^3 \Delta\theta}$	$\underline{Fo} = \dfrac{\pi_c}{\pi_k}$	Baron Jean Baptist Joseph Fourier; French mathematician and thermodynamicist; 1768-1830
$\pi_i = \dfrac{F}{\rho l^2 v^2}$ $\pi_g = \dfrac{F}{\rho g l^3}$	$\underline{Fr} = \sqrt{\dfrac{\pi_g}{\pi_i}}$	William Froude; British naval architect; 1810-1879

Pi-Numbers

(1) Name	(2) Symbol	(3) Product Derived from (6) and (7)	(4) Fields of Application	(5) Constituting Laws, Equations, and Definitions
Galileo	$\underline{Ga}$ $=\left(\dfrac{Re}{Fr}\right)^2$	$\dfrac{g\,l^{\,3}}{\nu^{\,2}}$	Motions induced by buoyancy	See Reynolds and Froude numbers
Graetz[1]	$\underline{Gz}$	$\dfrac{c_p\,\dot{m}\;\text{gas}}{k\,l}$	Heat acquisition of cold gas flowing through heated pipe	Same as Fourier number. Relaxation: heat is assumed to be *conducted* through thermal boundary layer of pipe wall.
Grashof[1]	$\underline{Gr}$	$\dfrac{g\,\beta\,l^3\Delta\theta}{\nu^{\,2}}$	Motions induced by buoyancy of heated fluid	Same as Galileo, plus • Law of thermal gas expansion $V_{hot}-V_{cold}=\beta\Delta\theta\,V_{cold}$; then, buoyancy force $F_t=g\,m_{cold}\,\beta\,\Delta\theta$
Hartmann	$\underline{M}$ (or $\underline{G}$ or $\underline{M}_M$)	$Bl\sqrt{\dfrac{\sigma}{\nu\rho}}$	Magnetohydro-dynamics	• Lorentz's equation for magnetic force $\vec{F}=g\,\vec{\nu}\times\vec{B}$ • Maxwell's equation of current induction $\vec{\nabla}\times\vec{E}=-\partial\vec{B}/\partial t$ • Ohm's law $\vec{j}=\sigma\vec{E}$ • Definition $\vec{j}=\rho_v\,\vec{\nu}$ • Shear stress of Newtonian fluid $\tau_{xy}=$ $\nu\rho\left(\dfrac{\partial\nu_x}{\partial y}+\dfrac{\partial\nu_y}{\partial x}\right)$, etc.

Named Pi-Numbers, their Definitions, Fields of Application, and Constituting Laws (Cont.)

(6) Constituting Pi-Numbers Derived from (5)	(7) Basic Definition of Named Pi-Number	(8) Biographical Data, References, and Remarks
See Reynolds and Froude numbers	$$Ga = \frac{\pi_v^{\,2}}{\pi_g \, \pi_i}$$	Galileo Galilei; Italian astronomer and physicist; 1564-1642
Same as Fourier number. Relaxation: heat is assumed to be *conducted* through thermal boundary layer of pipe wall.	$$Gz = \frac{\pi_k}{\pi_c} \quad ^{(1)}$$ with $\dfrac{\rho \ell^3}{t} \triangleq \dot{m}_{gas}$	Leo Graetz; German physicist; 1856-1941 [1]Special form of Fourier number
$$\pi_f = \frac{F}{g \, \rho \, \ell^3 \beta \Delta \theta}$$ For π_v and π_i, see $\underline{Re}$	$$Gr = \frac{\pi_v^{\,2}}{\pi_f \, \pi_i} \quad ^{(1)}$$	Franz Grashof; German research engineer; 1826-1893 [1]Special form of Galileo number
$$\pi_{m2} = \frac{F}{\sigma \, \ell^3 B^2 v}$$ (Note 3) $$\pi_v = \frac{F}{v \rho \ell v}$$	$$M = \sqrt{\frac{\pi_v}{\pi_{m2}}}$$	Julius Frederic Georg Paul Hartmann; Danish physicist; 1881-1951

(1) Name	(2) Symbol	(3) Product Derived from (6) and (7)	(4) Fields of Application	(5) Constituting Laws, Equations, and Definitions
Hedstrom	He	$\dfrac{\rho\, t_0\, \ell^2}{\mu_B^2}$	Rapid flow of rigid-viscoplastic materials	See Bingham number, plus • Newton's law of inertia $\vec{F} = m\,\vec{a}$
Knudsen	Kn (1)	$\dfrac{\bar{\lambda}}{\ell}$	Flow of rarefied gases past wall	• Absolute viscosity of rarefied gases $\mu_r = \rho\,\bar{\lambda}\,a_s\sqrt{2/(\pi\gamma)}$ • Shear stress of gas $\tau = \mu_r\,\partial v/\partial y$ • Adiabatic compression of ideal gas $p = \dfrac{\rho}{\gamma}\,a_s^2$ • Newton's law of inertia $\vec{F} = m\,\vec{a}$
Lagrange	La	$\dfrac{p\,\ell}{\mu\,v}$	Laminar flow	• Shear stress of Newtonian fluid $\tau_{xy} =$ $\mu\left(\partial v_x/\partial y + \partial v_y/\partial x\right)$, etc.
Lewis	$Le = Pe^{*}\,Fo$	$\dfrac{\alpha}{D}$ (1)	Combined heat and mass transfer	See Fourier and Peclet mass transfer numbers
Lift[1] coefficient	c_L	$\dfrac{2\,F_L}{\rho\,v^2 A}$ (2)	Lift of bodies moving through fluid	See Newton number (with $F = F_L$ = lift force)
M	M $= \dfrac{We^3}{Fr^2\,Re^4}$	$\dfrac{g\,\mu^4}{\rho\,\sigma^3}$ (1)	Bubble motions in liquids	See Weber, Froude, and Reynolds numbers

Named Pi-Numbers, their Definitions, Fields of Application, and Constituting Laws (Cont.)

(6) Constituting Pi-Numbers Derived from (5)	(7) Basic Definition of Named Pi-Number	(8) Biographical Data, References, and Remarks
$\pi_i = \dfrac{F}{\rho \ell^2 v^2}$ For π_o and π_{VB}, see $\underline{Bm}$	$\underline{He} = \dfrac{\pi_{VB}^2}{\pi_o\,\pi_i}$ with $F \triangleq \tau\ell^2$	Bengt Olof Arvid Hedström; Swedish chemical engineer; born 1926
$\mu_r \triangleq \dfrac{\rho\,\bar{\lambda}\,a_s}{\sqrt{\gamma}}$ $\tau \triangleq \mu_r\,\dfrac{v}{\ell} \longrightarrow$ $F \triangleq \tau\ell^2 \longrightarrow \pi_{Vr} = \dfrac{F\sqrt{\gamma}}{\ell v \rho\,\bar{\lambda}\,a_s}$ $p \triangleq \dfrac{\rho}{\gamma}\,a_s^2$ $F \triangleq p\ell^2 \longrightarrow \pi_p = \dfrac{F\gamma}{\ell^2 \rho\,a_s^2}$ $\pi_i = \dfrac{F}{\rho\ell^2 v^2}$	$\underline{Kn} = \dfrac{\sqrt{\pi_p\,\pi_i}}{\pi_{Vr}}$	Martin Hans Christian Knudsen; Danish physicist; 1871-1949 [1] $\underline{Kn} = \dfrac{Ma}{Re}\,\sqrt{\gamma}$ for rarefied gases
$\pi_V = \dfrac{F}{\mu \ell v}$	$\underline{La} = \pi_V$ with $F \triangleq p\ell^2$	Comte Joseph Louis Lagrange; French geometer and astronomer; 1736-1813
See Fourier and Peclet mass transfer numbers		Warren Kendall Lewis; American chemical engineer; 1882-1975 [1]A material property
See Newton number (with $F = F_L$ = lift force)		[1]Special form of Newton number [2]The factor of "2" has no physical significance
See Weber, Froude, and Reynolds numbers	$M = \dfrac{\pi_s^3\,\pi_i^2}{\pi_g\,\pi_V^4}$ with $E \triangleq F\ell$	J. Fluid Mech. 23, Pt. 4, 749-766 (1965) NASA TND-5462, Oct. 1969 [1]A material property

(1) Name	(2) Symbol	(3) Product Derived from (6) and (7)	(4) Fields of Application	(5) Constituting Laws, Equations, and Definitions
Mach	$\underline{M}a$	$\dfrac{v}{a}$	Rapid compression of gas; supersonic gas flow	• Adiabatic compression of ideal gas $p = \dfrac{\rho}{\gamma}\, a_s^2$ • Newton's law of inertia $\vec{F} = m\,\vec{a}$
Magnetic[1] Mach	$\underline{M}m = \underline{AL}$	$\dfrac{v}{v_A}$	Magnetohydro-dynamics	• Alfven speed $v_A = H\sqrt{\mu/\rho}$ introduced into Alfven number
Magnetic[1] Prandtl	$\underline{Pr}_m$ $= \underline{R}_m/\underline{R}e$	$\dfrac{\nu}{\eta}$ [1]	Magnetohydro-dynamics	See Reynolds and Magnetic Reynolds number
Magnetic pressure	$\underline{R}_H$ $= (\underline{AL})^2$	$\dfrac{\mu H^2}{\rho v^2}$ [1]	Magnetohydro-dynamics	See Alfven number
Magnetic[1] Reynolds	$\underline{R}_m$ (or $\underline{R}_M$, or $\underline{R}_v$)	$\dfrac{\ell v}{\eta}$ [1]	Magnetohydro-dynamics	• Lorentz's force equations Magnetic force $\vec{F} = g\,\vec{v} \times \vec{B}$ Electric force $\vec{F} = g\,\vec{E}$ • Maxwell's equation of magnetic induction $\vec{\nabla} \times \vec{H} = \vec{j} \quad (\vec{D} \approx 0)$ • Relation between $\vec{B}$ and $\vec{H}$ $\vec{B} = \mu\,\vec{H}$ • Ohm's law $\vec{j} = \sigma\,\vec{E}$

Named Pi-Numbers, their Definitions, Fields of Application, and Constituting Laws (Cont.)

(6) Constituting Pi-Numbers Derived from (5)	(7) Basic Definition of Named Pi-Number	(8) Biographical Data, References, and Remarks
$F \triangleq \rho \ell^2 \longrightarrow \pi_p = \dfrac{F\gamma}{\rho a_s^2 \ell^2}$ $p \triangleq \dfrac{\rho}{\gamma} a_s^2$ $\pi_i = \dfrac{F}{\rho \ell^2 v^2}$	$Ma = \sqrt{\dfrac{\pi_p}{\pi_i \gamma}}$	Ernst Mach; German physicist; 1838–1916
Alfven speed $v_A = H\sqrt{\mu/\rho}$ introduced into Alfven number		[1] A special form of the Alfven number
See Reynolds and Magnetic Reynolds numbers		[1] A material property analogous to the Prandtl number
See Alfven number		[1] ρv^2 represents dynamic pressure; μH^2, magnetic pressure
$\pi_{m3} = \dfrac{F}{g \, v \, \mu \, j \, \ell}$ $\pi_{e\ell} = \dfrac{F\sigma}{g \, j}$ (Note 4)	$R_m = \dfrac{\pi_{e\ell}}{\pi_{m3}}$	[1] Structure of R_m analogous to Re. Hence the name.

S.M.E.—K

(1) Name	(2) Symbol	(3) Product Derived from (6) and (7)	(4) Fields of Application	(5) Constituting Laws, Equations, and Definitions
McAdams	—	$\dfrac{h^4 \ell \mu \Delta\theta}{k^3 \rho^2 g \lambda}$	Heat transfer from condensating vapor to cold surface	See Condensation number
Newton	$\underline{Ne}$	$\dfrac{F}{\rho \ell^2 v^2}$	Phenomena involving inertial forces	• Newton's law of inertia $\vec{F} = m\vec{a}$
Nusselt	$\underline{Nu}$	$\dfrac{h\ell}{k}$ (1)	Heat convection	• Definition of heat flux between fluid and wall $g = h\left(\theta_{\text{surf}} - \theta_{\text{fluid}}\right)$ • Fourier's law of heat conduction $g = -k\,\text{grad}\,\theta$
Ohnesorge	$\underline{Z}$ $= \dfrac{\sqrt{We}}{Re}$	$\dfrac{\mu}{\sqrt{\rho \sigma \ell}}$	Atomization of fluids	See Weber and Reynolds numbers
Peclet	$\underline{Pe}$ $= 1/\underline{Fo}$	$\dfrac{\ell v}{\alpha}$	Heat conduction	See Fourier number
Peclet number for mass transfer	$\underline{Pe}^*$	$\dfrac{\ell v}{D}$	Mass transfer by diffusion	• Fick's law of diffusion $j = -D\,\text{grad}\,\rho$
Prandtl	$\underline{Pr}$ $= \left(\underline{Re}\ \underline{Fo}\right)^{-1}$	$\dfrac{\nu}{\alpha}$ (1)	Forced heat convection	See Reynolds and Fourier numbers

Named Pi-Numbers, their Definitions, Fields of Application, and Constituting Laws (Cont.)

(6) Constituting Pi-Numbers Derived from (5)	(7) Basic Definition of Named Pi-Number	(8) Biographical Data, References, and Remarks
See Condensation number	McAdams Group $= \dfrac{\pi_k^3 \, \pi_g \, \pi_\lambda}{\pi_h^4 \, \pi_v}$	William Henry McAdams; American chemical engineer; 1892-1975
$\pi_i = \dfrac{F}{\rho \, l^2 \, v^2}$	$\underline{Ne} = \pi_i$	Sir Isaac Newton; British mathematician and natural philosopher; 1642-1727
$\pi_h = \dfrac{Q}{h \, l^2 \, \Delta \theta t}$ $\pi_k = \dfrac{Q}{k \, l \, \Delta \theta t}$	$\underline{Nu} = \dfrac{\pi_k}{\pi_h}$	Wilhelm Nusselt; German thermodynamicist; 1882-1957 [1]See Note 2
See Weber and Reynolds numbers		Wolfgang von Ohnesorge; German fluid dynamicist; see *Zeitschrift für Angewandte Mathematik und Mechanik* 16, 355-358 (1936)
See Fourier number		Jean-Claude Eugène Péclet; French physicist; 1793-1857
$\pi_D = \dfrac{m}{D \rho \, l t}$	$\underline{Pe}^* = \pi_D$ with $m \triangleq \rho \, l^3$	
		Ludwig Prandtl; German aerodynamicist; 1875-1953 [1]A material property

(1) Name	(2) Symbol	(3) Product Derived from (6) and (7)	(4) Fields of Application	(5) Constituting Laws, Equations, and Definitions
Rayleigh	Ra $= Gr\,Pr$	$\dfrac{g\,l^3\beta\Delta\Theta}{\alpha\,\nu}$	Heat convection in gases	See Grashof and Prandtl numbers
Reynolds	Re	$\dfrac{l\,v}{\nu}$	Turbulent flow	• Shear stress of Newtonian fluid $\tau_{xy}=\mu\left(\dfrac{\partial v_x}{\partial y}+\dfrac{\partial v_y}{\partial x}\right)$; etc. • Newton's law of inertia $\vec{F}=m\,\vec{a}$
Richardson[1]	Ri	$\dfrac{g\,(\partial\rho/\partial z)}{\rho\,(\partial\bar{v}/\partial z)^2}$	Vertical motions of thermally stratified fluids	See Note 5
Rossby	Ro	$\dfrac{v}{\omega\,l}$	Large-scale atmospheric or oceanic motions	• Newton's law of inertia $\vec{F}=m\,\vec{a}$ Relaxation: Inertial forces separated into centrifugal and Coriolis components $\vec{F}_{ce}=m\,\vec{\omega}\times\left(\vec{\omega}\times\vec{r}\right)$ $\vec{F}_{co}=2\,m\,\vec{\omega}\times\vec{v}$
Schmidt	Sc $=\dfrac{Pe^*}{Re}$	(1) $\dfrac{\nu}{D}$	Flow with momentum and mass transfer	See Peclet number for mass transfer and Reynolds number
Sherwood	Sh	$\dfrac{h_D\,l}{D}$	Mass transfer by convection	• Definition of mass transfer coefficient $h_D=j/\Delta\rho$ • Fick's law of diffusion $j=-D\,\mathrm{grad}\,\rho$

Named Pi-Numbers, their Definitions, Fields of Application, and Constituting Laws (Cont.)

(6) Constituting Pi-Numbers Derived from (5)	(7) Basic Definition of Named Pi-Number	(8) Biographical Data, References, and Remarks
See Grashof and Prandtl numbers		John William Strutt Third Baron Rayleigh; British physicist; 1842-1919
$\pi_v = \dfrac{F}{\mu l v}$ $\pi_i = \dfrac{F}{\rho l^2 v^2}$	$Re = \dfrac{\pi_v}{\pi_i}$	Osborne Reynolds; British physicist; 1842-1912
See Note 5		Lewis Fry Richardson; British physicist and meteorologist; 1881-1953 [1]Special form of Froude number
$\pi_{ce} = \dfrac{F}{m l \omega^2}$ $\pi_{co} = \dfrac{F}{m \omega v}$	$Ro = \dfrac{\pi_{ce}}{\pi_{co}}$	Carl Gustav (Arvid) Rossby; Swedish/US meteorologist; 1898-1957
See Peclet number for mass transfer and Reynolds number		Ernst Heinrich Wilhelm Schmidt; German thermodynamicist; 1892-1975 [1]A material property
$\pi_{hm} = \dfrac{m}{h_D \rho l^2 t}$ $\pi_D = \dfrac{m}{D l \rho t}$	$Sh = \dfrac{\pi_D}{\pi_{hm}}$	Thomas Kilgore Sherwood; American chemical engineer; 1903-1976

(1) Name	(2) Symbol	(3) Product Derived from (6) and (7)	(4) Fields of Application	(5) Constituting Laws, Equations, and Definitions
Sommerfeld[1]	$\underline{So}$	$$\frac{\mu N}{p\,(c/r)^2}$$ c/r = clearance modulus	Friction in journal bearings	See Note 6
Specific[1] speed	$\underline{N_s}$ $$=\left(\frac{2}{Eu}\right)^{3/4}$$	$$\frac{N\sqrt{P/\rho}}{\sqrt[4]{(Hg)^5}} \quad ^{(2)}$$	Hydro-dynamic machinery	See Euler number
Stanton	$\underline{St}$ $= \underline{Nu}\,\underline{Fo}$	$$\frac{h}{c_p \rho v}$$	Heat convection between fluid and wall	See Nusselt and Fourier numbers
Stefan	—	$$\frac{e\,\sigma_0\,\theta^3 l}{k}$$	Heat transfer through radiation	• Stefan-Boltzmann's law of radiated energy emitted by grey body $$g = e\,\sigma_0\,\theta^4$$ • Fourier's law of heat conduction $$g = -k\ \mathrm{grad}\ \theta$$
Strouhal[1]	$\underline{St}$	$$\frac{fl}{v}$$ $$f = \sqrt{\frac{k}{m}}$$ natural frequency	Elastic vibrations; vortex shedding[2]	See Cauchy number. Relaxations: All elastic properties of system lumped into gross spring rate, k. All masses lumped into gross mass, m.

Named Pi-Numbers, their Definitions, Fields of Application, and Constituting Laws (Cont.)

(6) Constituting Pi-Numbers Derived from (5)	(7) Basic Definition of Named Pi-Number	(8) Biographical Data, References, and Remarks
See Note 6		Arnold Johannes Wilhelm Sommerfeld; German physicist; 1868-1951 [1]A special form of the Lagrange number
To adapt the Euler number to the nomenclature in hydro-dynamic machinery, a number of substitutions are made: $$\left(\frac{2}{Eu}\right)^{3/4} \longrightarrow N_s$$ $p \triangleq \rho g H$ $v \triangleq \ell/t$ $\ell \triangleq \sqrt[3]{\dot{V}t}$ $t \triangleq 1/N$ $P \triangleq \rho g \dot{V} H \longrightarrow \dot{V} \triangleq \dfrac{P}{\rho g H}$		[1]A special form of the Euler number [2]The gravitational acceleration, g , is usually omitted so that, unfortunately, N_s loses its dimensionless character. Sir Thomas Ernest Stanton; British civil and mechanical engineer; 1865-1931
$\pi_r = \dfrac{Q}{e\,\sigma_o\,\ell^2 t\,\theta^4}$ $\pi_k = \dfrac{Q}{k\,\ell\,\theta\,t}$	Stefan number $= \dfrac{\pi_k}{\pi_r}$	Josef Stefan; Austrian physicist; 1835-1893
$\pi_{es} = \dfrac{F}{k\,\ell}$ $\pi_i = \dfrac{F\ell}{m\,v^2}$	$St = \sqrt{\dfrac{\pi_i}{\pi_{es}}}$	Vincenz Strouhal; Czech physicist; 1850-1922 [1]A special form of the Cauchy number [2]See Note 7

See Section "Segmented Modeling," Part I

(1) Name	(2) Symbol	(3) Product Derived from (6) and (7)	(4) Fields of Application	(5) Constituting Laws, Equations, and Definitions
Taylor[1]	$\underline{Ta}$	$\dfrac{\omega \sqrt{R_a b^3}}{\nu}$ R_a = mean radius of annulus b = width of annulus	Flow between rotating concentric cylinders	Same as Dean number with $R = R_a , r = b$, and $\nu = R_a \omega$ See Section "Segmented Modeling," Part I.
Thoma	$\underline{\sigma}$ (1)	$\dfrac{p_h - p_v}{p_2 - p_1}$	Cavitation	• Difference between total hydrostatic pressure at point of interest, p_h , and vapor pressure at given temperature, p_v . • Difference between absolute pressure at delivery side, p_1 , and at inlet side, p_2 .
Thring	$\underline{Th}$ $= \dfrac{1}{\underline{Fo} \times \text{Stefan}}$	$\dfrac{c_p \rho v}{e \sigma \theta^3}$	Heat transfer through radiation	See Fourier and Stefan numbers
Velocity[1] ratio	u^+	$\dfrac{\bar{u}}{u_\tau}$	Flow past wall	• Newton's law of inertia $\vec{F} = m \vec{a}$ Relaxation: In turbulent flow along a wall, the inertial force at some distance from the wall may be expressed in terms of the time average of the velocity parallel to the wall, $\bar{u}$.
Weber	$\underline{We}$	$\dfrac{\rho l v^2}{\sigma}$	Surface tension effects	• Surface energy of liquids $E = \sigma A$ • Kinetic energy (Newton's law of inertia) $E = \dfrac{m}{2} v^2$
Z		Identical with Ohnesorge number		

Named Pi-Numbers, their Definitions, Fields of Application, and Constituting Laws (Concl.)

(6) Constituting Pi-Numbers Derived from (5)	(7) Basic Definition of Named Pi-Number	(8) Biographical Data, References, and Remarks
Same as Dean number with $R = R_a$, $r = \ell$, and $v = R_a \omega$ See Section "Segmented Modeling," Part I.		Sir Geoffrey Ingram Taylor; British meteorologist and physicist; 1886-1975 [1]A special form of the Reynolds number
• Difference between total hydrostatic pressure at point of interest, p_h, and vapor pressure at given temperature, p_v. • Difference between absolute pressure at delivery side, p_1, and at inlet side, p_2.		Dieter Thoma; German hydraulic engineer; 1881-1942 [1] σ is not a principal pi-number
See Fourier and Stefan numbers		Meredith Wooldridge Thring; British chemical engineer; born 1915
$F \triangleq \rho \, \ell^2 \bar{u}^2 \longrightarrow \Pi_w = \dfrac{\tau_w}{\rho \, \bar{u}^2}$ $F \triangleq \tau_w \, \ell^2$ τ_w = actual wall shear stress	$u^+ = \dfrac{1}{\sqrt{\Pi_w}}$, with "friction velocity" $u_\tau \equiv \sqrt{\tau_w / \rho}$	[1]A special form of the Newton number
$\Pi_s = \dfrac{E}{\sigma \ell^2}$ $\Pi_i = \dfrac{E}{\rho \, \ell^3 v^2}$	$\underline{We} = \dfrac{\Pi_s}{\Pi_i}$	Moritz Weber; German research engineer; 1871-1951
	Identical with Ohnesorge number	

Notes on the Catalog of Principal Pi-numbers

Note 1

The two definitions given in column (5) for electric current density, j, and electric current, i, combined with definitions of electric charge density, ρ_V, and velocity, v, lead to the following representative expression for electric charge, q,

$$i \triangleq j\ell^2 \longrightarrow i \triangleq \ell^2 \rho_V v \longrightarrow q \triangleq it$$

$$j \triangleq \rho_V v \qquad\qquad \rho_V \triangleq \frac{q}{\ell^3} \qquad v \triangleq \frac{\ell}{t}$$

The given Maxwell equation can be representatively expressed as

$$H \triangleq j\ell \longrightarrow i \triangleq H\ell$$

$$j \triangleq \frac{i}{\ell^2}$$

With these representative expressions, and with the relation between B and H, column (5), Lorentz's equation for magnetic force can be transformed into the expression given in column (6),

$$F \triangleq q v B \longrightarrow F \triangleq H^2 \ell^2 \mu \longrightarrow \pi_{m1} = \frac{F}{H^2 \ell^2 \mu}$$

$$q \triangleq it \qquad v \triangleq \frac{\ell}{t} \qquad i \triangleq H\ell \qquad B \triangleq \mu H$$

Note 2

The Nusselt number is traditionally restricted to heat phenomena in a fluid, so that k is always taken as the thermal conductivity of a fluid, and h, the surface coefficient of heat transfer, as an unknown quantity that must be determined experimentally.

The Biot number, by contrast, it traditionally applied to heat transfer problems within solid bodies so that k is taken as the thermal conductivity of the body; h is considered a given quantity at the body's boundaries. In form, $\underline{Nu}$ and $\underline{Bi}$ are identical.

Note 3

The definition for electric current density, j, given in column (5) combined
with the definition for electric charge density, ρ_V, leads to a representative
expression for electric charge, q,

$$j \triangleq \rho_V v \longrightarrow q = \frac{j \ell^3}{v}$$

$$\rho_V \triangleq \frac{q}{\ell^3}$$

With this expression, and with representative forms of Ohm's law and Maxwell's
equation, both given in column (5), Lorentz's equation for magnetic force
can be transformed in the representative expression quoted in column (6)

$$F \triangleq q v B \longrightarrow F \triangleq \sigma \ell^3 B^2 v \longrightarrow \pi_{m2} = \frac{F}{\sigma \ell^3 B^2 v}$$

$$q \triangleq \frac{j \ell^3}{v} \qquad j \triangleq \sigma E \qquad E \triangleq B \frac{\ell}{t}$$

Note 4

A third representative version of the magnetic force (see Notes 1 and 3) is
derived from the laws given in column (5) as

$$F \triangleq q v B \longrightarrow F \triangleq q v \mu j \ell \longrightarrow \pi_{m3} = \frac{F}{q v \mu j \ell}$$

$$H \triangleq j \ell \qquad B \triangleq \mu H$$

The representative electric force and the equivalent principal pi-number
quoted in column (6) derive from the Lorentz force equation and Ohm's law.

$$F \triangleq q E \longrightarrow F \triangleq q \frac{j}{\sigma} \longrightarrow \pi_{el} = \frac{F \sigma}{q j}$$

$$E \triangleq \frac{j}{\sigma}$$

Note 5

The Richardson number is, like the Froude number, based on Newton's laws of
inertia and gravitation. Its application, however, is mostly restricted to
the vertical turbulence of a horizontally flowing fluid with vertical
velocity and density gradients, $\partial \bar{v} / \partial z$ and $\partial \rho / \partial z$, respectively, where z is
the vertical direction. Vertical turbulence can be visualized as caused by

vertical migrations (up and down) of small fluid lumps, with a mean absolute velocity of $|w''|$. The kinetic energy of these lumps is representatively described by $E \triangleq \rho V\, w''^2$, where V represents the volume of the fluid lumps; and ρ, the fluid density.

At this point, a basic assumption well established in boundary layer theory is introduced: the mean vertical migration velocity, $|w''|$, is considered proportional to the difference between the horizontal velocities of two closely spaced fluid layers, Δv; i.e., $|w''| \propto \Delta v$. (The distance between the two layers is known as Prandtl's mixing length; it is the mean distance a fluid lump travels upward or downward before it loses its identity.) Since, in representative terms, $\Delta v \triangleq z\, \dfrac{\partial \bar{v}}{\partial z}$, $|w''|$ can also be expressed as $w'' \triangleq z\, \dfrac{\partial \bar{v}}{\partial z}$, where $\bar{v}$ is the mean horizontal velocity of a representative layer. This expression, substituted into the expression for E, yields the final form of the representative kinetic energy (in the vertical direction) $E \triangleq \rho V z^2 \left(\partial \bar{v}/\partial z\right)^2$. Consequently,

$$\pi_{km} = \frac{E}{\rho V z^2 (\partial \bar{v}/\partial z)^2} .$$

The buoyancy energy of a fluid lump is represented by $E \triangleq \Delta \rho\, g V z$, where $\Delta \rho$ is the representative difference between the densities of a displaced lump and its environment. Since $\Delta \rho$ can be expressed by the density gradient, $\Delta \rho \triangleq z\, (\partial \rho/\partial z)$, we arrive at the expression for representative buoyancy energy of $E \triangleq z^2 g V (\partial \rho/\partial z)$ and, hence, at $\pi_{bm} = \dfrac{E}{z^2 g V (\partial \rho/\partial z)}$.

The Richardson number is defined as $\underline{Ri} = \pi_{km}/\pi_{bm}$. Therefore,

$$\underline{Ri} = \frac{g\,(\partial \rho/\partial z)}{\rho\,(\partial \bar{v}/\partial z)^2}$$

Note 6

The Sommerfeld number[1] is a special case of the Lagrange number applied to friction in journal bearings. The oil film between shaft and journal bearing is exposed to two forces -- pressure and viscosity. As explained in the

---

[1]D.D. Fuller, *Theory and Practice of Lubrication for Engineers*, Wiley, London, 1956.

section "Segmented Modeling," the two forces act in different regions: viscous
forces are restricted to surface areas of the shaft and bearing, $A_S \triangleq rL$;
whereas pressure forces act on the cross-sectional area, $A_C = cL$, of the oil
film, where r is the shaft radius, L is the axial length of the bearing, and
c is the oil film thickness.

Accordingly,

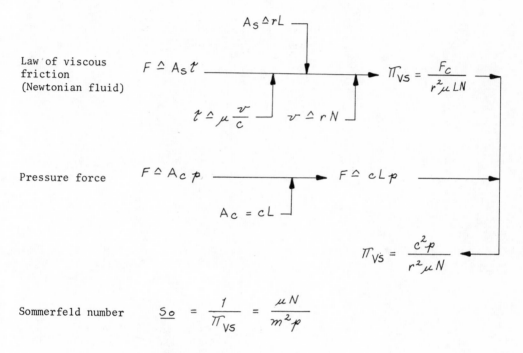

Law of viscous friction (Newtonian fluid)

Pressure force

Sommerfeld number

where N is the number of revolutions per unit time, and $m = c/r$ is the clear-
ance modulus.

Note 7

When an elastic structure is immersed in a stream of air or water, periodic
vortex shedding may occur, called Karman vortex street (behind transmission
lines, bridges, airfoils). The frequency of vortex formation depends on the
fluid speed. If the frequency of the vortex formation coincides with the
natural frequency of the structure, violent vibrations may occur. Often, the
Strouhal number is used to identify this situation. Here, ν represents the

wind velocity, f the vortex shedding frequency, and ℓ the width of the cross section. The Strouhal number in this case is a dimensionless vortex frequency whose numerical value should be kept at a safe distance from the numerical value of the Strouhal number computed from the vibrational characteristics of the structure.

APPENDIX C

NOTATION

a	acceleration	m/s^2
a, a_s	speed of sound	m/s^2
A	area	m^2
B	magnetic induction (or magnetic flux density)	T
c	coefficient of soil cohesion	N/m^2
c_p	specific heat at constant pressure	J/(kg K)
c_v	specific heat at constant volume	J/(kg K)
D	diffusivity	m^2/s
D	electric displacement	C/m^2
e	emissivity	-
E	energy	J
E	electric field strength	V/m
E	Young's modulus	N/m^2
f	frequency	1/s or Hz
F	force	N
F_D	drag force	N
F_L	lift force	N
g	gravitational acceleration	m/s^2
h	surface coefficient of heat transfer	$J/(m^2 \text{ s K})$
h_D	mass transfer coefficient	m/s
H	magnetic field strength	A/m
H	hydraulic head	m
i	electric current	A
j	mass flux	$kg/(m^2 \text{ s})$
j	electric current density	A/m^2
k	thermal conductivity	J/(m s K)
k	permeability (of porous media)	m^2
ℓ	length	m
m	mass	kg
N	number of revolutions in unit time	1/s
p	pressure	N/m^2 or Pa
P	power	W

283

q	energy flux, heat flux	W/m^2
q	electric charge	C
q_e	exothermic heat per unit mass	J/kg
Q	heat energy	J
R	gas constant	$m^2/(s^2\ K)$
t	time	s
U	reaction rate	$kg/(m^3\ s)$
v	velocity	m/s
V	volume	m^3

$\alpha \equiv \dfrac{k}{c_p \rho}$	thermal diffusivity	m^2/s
β	coefficient of thermal expansion	1/K
$\gamma \equiv \dfrac{c_p}{c_v}$		-
γ	shear strain	-
ϵ	normal strain	-
$\eta \equiv \dfrac{1}{\sigma \mu}$	magnetic diffusivity (σ = el. cond., μ = mag. perm.)	m^2/s
θ	temperature	K
λ	latent heat of condensation per unit mass	J/kg
$\bar{\lambda}$	mean free path length of molecules	m
μ	coefficient of (internal soil) friction	-
μ	dynamic (absolute) viscosity	$N\ s/m^2$
μ	magnetic permeability	Wb/(A m)
μ_r	dynamic viscosity of rarefied gas	$N\ s/m^2$
$\nu \equiv \dfrac{\mu}{\rho}$	kinematic viscosity (μ = dyn. visc.)	m^2/s
ν	Poisson ratio	-
ρ	mass density	kg/m^3
ρ_v	electric charge density	C/m^3
σ	normal stress	N/m^2
σ, σ_s	surface tension	N/m
σ_o	Stefan-Boltzmann constant	$W/(m^2\ K^4)$
σ	electric conductivity	$1/(\Omega\ m)$
τ	shear stress	N/m^2
ω	angular velocity	1/s

APPENDIX D

SI AND OTHER UNITS

TABLE D.1. Units of Primary Quantities.

PRIMARY QUANTITY	SI UNIT	OTHER UNITS
LENGTH	m (meter or metre)	angstrom (A) = 10^{-10} m foot (ft) = 0.3048 m
TIME	s (second)	
FORCE	N (newton)	dyne = 10^{-5} N pound force (lb) = 4.448 N kilogram force (kgf) = kilopond (kp) = 9.807 N
TEMPERATURE	K (kelvin)	degree Celsius (^{O}C) = K - 273.15 degree Fahrenheit (^{O}F) = 1.8 K - 459.67 degree Rankine (^{O}R) = 1.8 K degree centigrade = ^{O}C
ELECTRIC CURRENT (CONDUCTION AND DISPLACEMENT)	A (ampere)	

Table D.2. Units of Some Secondary Quantities.

SECONDARY QUANTITY	SI UNIT	OTHER UNITS
FREQUENCY	hertz (Hz) = 1/s	
MASS*	kilogram (kg) = N s^2/m	slug = lb s^2/ft = 14.594 kg
WORK, ENERGY, HEAT	joule (J) = N m	calorie (int'l) = 4.1868 J erg = 10^{-7} J British thermal unit (Btu, int'l) = 1055.06 J
POWER	watt (W) = J/s	horsepower (hp) = 550 ft lb/s = 745.7 W
PRESSURE	pascal (Pa) = N/m^2	bar = 10^5 Pa atmosphere (techn) = 0.9807 x 10^5 Pa torr = 133.3 Pa
DYNAMIC VISCOSITY	Pa s	centipoise (cp) = 10^{-3} Pa s
KINEMATIC VISCOSITY	m^2/s	centistokes (cs) = 10^{-6} m^2/s
ELECTRIC CHARGE	coulomb (C) = A s	
ELECTRIC POTENTIAL	volt (V) = W/A	
ELECTRIC RESISTANCE	ohm (Ω) = V/A	
ELECTRIC CAPACITANCE	farad (F) = C/V	
ELECTRIC CONDUCTANCE	siemens (S) = A/V	mho = S
MAGNETIC FLUX	weber (Wb) = V s	maxwell = 10^{-8} Wb
ELECTRIC INDUCTANCE	henry (H) = Wb/A	
MAGNETIC FLUX DENSITY	tesla (T) = Wb/m^2	gauss = 10^{-4} T

*IN THE SI SYSTEM, MASS IS A PRIMARY QUANTITY. HERE, WE CONSIDER MASS A SECONDARY
QUANTITY DERIVED FROM THE PRIMARY QUANTITY OF FORCE.

APPENDIX E
ANSWERS TO PROBLEMS

1. Frequency scale factor $f^* = a_s^*/\ell^*$

   Force scale factor $F^* = E^*\ell^{*2}$

2. The governing pi-number, $\pi = \rho g \ell/E$, indicates that if the same materials are used for both model and prototype, the length scale factor is unity. Hence, model tests are impossible, unless relaxations such as dummy weights are introduced (see Chapter 3).

3. Both coefficients, e and μ, are principal pi-numbers within the range of validity of their constitutive equations. "Slip" and "efficiency" are common pi-numbers since they are based on definitions, not laws.

4. The Hodgson number is a common pi-number.

6. Model rule for pendulum mass $m^* = \ell^{*2}$

   Time scale factor $t^* = \sqrt{\ell^*}$

   Scale factor of final angle $\phi^* = 1$

7. All terms are representative expressions of a force per unit length. For the first and last term, the forces are inertial forces; for the third term, weight forces. Hence, the Froude number applies.

8. The Euler number assumes the form $\pi = \rho \ell^2/\dot{m}\sqrt{R\theta}$.

9. Scale factor of sloshing frequency $f^* = \sqrt{a^*/\ell^*}$

10. The Newton number can be expressed as $\pi = R\theta / \ell^2 N^2$

    Scale factor of rotational speed $N^* = \sqrt{R^*\theta^*} / \ell^*$

    Scale factor of mass flow rate $\dot{m}^* = \rho^* \ell^{*2} / \sqrt{R^*\theta^*}$

    Scale factor of specific work $h^* = R^*\theta^*$

    Scale factor of torque $T^* = \rho^* \ell^{*3}$

11. Speed scale factor $v^* = 1$

    Frequency scale factor $f^* = 0.05$

    Wing mass scale factor $m^* = 8000$

    Corresponding speeds and angles of model and prototype are equal.

12. Elastic forces of a blood cell are governed by $\pi_e = F / E\ell^2$

    Viscous forces of both intracellular and extracellular fluids are governed by $\pi_v = F / \mu \ell v$.

    Young's modulus of a red blood cell $E = E'(\mu/\mu')(t'/t)$ $= 0.67 \times 10^5$, N/m^2

    The viscosity ratio of 4.1 applies to both model and prototype.

15. The density scale factor can be selected arbitrarily.

    Ship mass scale factor $m^* = \rho^* \ell^{*3}$ $= 10^6$

    Radius of gyration scale factor $= \ell^* = 100$

    Frequency scale factor $f^* = \sqrt{1 / \ell^*}$ $= 0.1$

16. Weight scale factor $W^* = k_g^* \ell^{*n+2}$ $= 4^{3.1} = 73.5$

17. Weight and drawbar pull scale factor $W^* = B^* = c^* \ell^{*2}$

    Power scale factor $P^* = c^* \ell^{*3}$

    Slip as a ratio of two speeds is a common pi-number; its value, like the values of all other pertinent common pi-numbers, must be kept the same for both model and prototype.

18. $v^* = 1$; $F^* = P^* = k^* = \ell^{*2} = 36$

19. $\ell^* = c_v^* \; \theta^* / g^* = 0.41 \; (672/85) \; 6 = 19.5$

$v^* = \sqrt{g^* \ell^*} = 1.8$

$F^* = \rho^* g^* \ell^{*3} = 1226$

20. With the pi-numbers π_i and π_g developed in Case Study 5, the dimensionless resistance used by Krause is π_i^3 / π_g^2.

Simpler expressions for the dimensionless force such as $F_H / \rho g \ell^3$ could have been used.

Comparisons at dimensionless depths of $t/t' \neq \ell^*$ would violate the fundamental principle of equal length scale factor for all pertinent lengths and distances.

22. $E^* = \ell^{*3}$

23. $p^* = 1$

The temperature scale factor of unity is enforced by starting both model and prototype tests at identical temperatures.

24. $E_o^* = \ell^*$; $n_o^* = 1/\ell^{*2}$; $T^* = \ell^*$; $\Delta\theta_{peak}^* = 1/\ell^{*2}$; $t_{stop}^* = \ell^{*2}$.

26. $A \triangleq m^{2/3}$; $V_L \triangleq m$; $f_L \triangleq m^{-1/3}$.

27. $E^* = \ell^{*3}$; $p^* = 1$.

APPENDIX F

INDEX

Z